Project Earth Science: Meteorology

Second Edition

NSTA Press, NSTA Journals, and the NSTA website deliver high-quality resources for science educators.

Project Earth Science: Meteorology
NSTA Stock Number: PB103X
ISBN 0-87355-123-0
Library of Congress Card Number: 95-067463
Printed in the USA by Reproductions, Inc.
Printed on recycled paper.
Cover design by Marty Ittner of AURAS Design, Inc.
Illustrations by Max-Karl Winkler.

The mission of the National Science Teachers Association is to promote excellence and innovation in science teaching and learning for all.

NSTA Press
1840 Wilson Boulevard
Arlington, Virginia 22201-3000
www.nsta.org

Project Earth Science: Meteorology

Second Edition

by

P. Sean Smith and Brent A. Ford

Now featuring *sci*LINKS®—a new way of connecting text and the Internet. Up-to-the-minute online content, classroom ideas, and other materials are just a click away. Go to page 214 to learn more about this educational resource.

Arlington, Virginia

A Project of Horizon Research, Inc.
Material for middle school teachers in Earth science
This project was funded by BP America, Inc.

Project Earth Science: Meteorology
Second Edition

Table of Contents

Activities

Readings

Appendices

About the Cover Photographs

Front Cover:

• National Weather Service (Patuxent River, MD, station) radar showing precipitation moving over Washington, DC, metropolitan area. Commonly shown on television weather broadcasts, radar images such as this are used extensively in forecasting and in tracking the movement of precipitation systems. Supplied by H. Michael Mogil.

• Close cloud-to-ground lightning flash in Norman, OK. The main channel is overexposed because it is illuminated several times during the course of the flash by successive strokes. Photographed by William Beasley.

Back Cover:

• Towering cumulus cloud with rain shaft (Barbados). Note the difference in weather below and above the tower. Also notice the shape of the tower; this is how strong thunderstorms commonly appear. Considerable lightning appeared in this cloud complex. Photographed by H. Michael Mogil.

• Perspective view of an incipient tornado within a thunderstorm simulated in three dimensions. The vertical tube indicates a region of strong rotary motion, and the two horizontal planes—at ground level and at 2 km above the ground—show storm-relative flow patterns (dark-lines) and temperature deviation from the environment. Blues indicate areas colder in the storm than in the surroundings, reds indicate warmer areas. Photographed by Kelvin Droegemeier and Ming Xue at the Center for Analysis and Prediction of Storms, University of Oklahoma.

Acknowledgments

Many people have contributed to *Project Earth Science: Meteorology*. The volume began as a collection of activities and readings for Project Earth Science, a teacher enhancement project funded by the National Science Foundation.

Project Earth Science was designed to provide inservice education for North Carolina middle school Earth science teachers. Nine two-person leadership teams received extensive training in conducting workshops on selected topics in astronomy, geology, meteorology, and oceanography. These teams, in turn, presented teacher enhancement programs throughout North Carolina. This book assembles what these teams developed. Principal investigators for this project were Iris R. Weiss, president of Horizon Research, Inc.; Diana Montgomery, research associate at Horizon Research, Inc.; Paul B. Hounshell, professor of education, University of North Carolina–Chapel Hill; and Paul Fullagar, professor of geology, University of North Carolina–Chapel Hill. Linda Ford, Novostar Designs, Inc., contributed significantly in the writing, editing, and illustrating of the readings as well as to developing a number of the activities.

The activities and readings have undergone many revisions following an extensive review process by project leaders, workshop participants, consultants, and project staff. Project leaders included: Kevin Barnard, Winston-Salem/Forsyth County Schools; Kathy Bobay, Charlotte-Mecklenburg Schools; Pam Bookout, Guilford County Schools; Betty Dean, Guilford County Schools; Lynanne (Missy) Gabriel, Charlotte-Mecklenburg Schools; Flo Gullickson, Guilford County Schools; Michele Heath, Chapel Hill/Carrboro Schools; Cameron Holbrook, Winston-Salem/Forsyth County Schools; Linda Hollingsworth, Randolph County Schools; Geoff Holt, formerly of Wake County Schools; Kim Kelly, formerly of Chapel Hill/Carrboro Schools; Laura Kolb, Wake County Schools; Karen Kozel, Durham County Schools; Kim Taylor, formerly of Durham County Schools; Dana White, Wake County Schools; Tammy Williams, Guilford County Schools; and Lowell Zeigler, Wake County Schools.

Project Earth Science: Meteorology was extensively reviewed for scientific accuracy. Steve Businger, professor of meteorology at North Carolina State University, and H. Michael Mogil, certified consulting meteorologist, How the Weather Works, contributed extensively to this review process. Parts or all of the manuscript were examined by Robert Hudson, professor of meteorology, University of Maryland College Park; Ira Geer, director of education programs, American Meteorological Society; Robert Weinbeck, associate professor of Earth sciences, State University of New York-Brockport; Donald McManor, American Meteorological Society; Chris Mastropieri and Toni DeVore, Project Atmosphere, American Meteorological Society; Gene Bierly, American Geophysical Union; Paul Mroz, Bergen, NY; Alfred Fowler, Council of Managers of National Antarctic Programs; and Ronald Holle, research meteorologist, National Severe Storms Laboratory, Norman, OK. Shirley Brown, Columbus City Schools, Ohio, provided much of the material in the appendix as well as many of the ideas for the "Suggestions for Interdisciplinary Reading and Study" sections within the activities.

Project Earth Science: Meteorology is published by the NSTA Press. The original edition was produced by Shirley Watt Ireton (director) and Andrew Saindon (project editor), with assistance from Gregg Sekscienski (associate editor); Elizabeth Cummings (editorial assistant); Joe Cain (associate editor); and Glen Fullmer (associate editor)

NSTA ordinarily uses only metric measurements in its publications. As of this writing, agencies of the United States government are in the process of converting from the English system of measurement to the metric system. Because this book makes use of and suggests using material supplied by government sources—and because the English system is standard in U.S. newspapers and television—we have included references to both systems here. Teachers are encouraged, however, to emphasize metric literacy in their science education.

Special thanks go to BP America, Inc. for providing the funds to make this book possible. For the second edition, thanks go to John Kermond, NOAA Office of Global Programs, and Erin Miller, Project Editor for NSTA Press.

Introduction

Project Earth Science: Meteorology is the second in the four-volume Project Earth Science series. The other three volumes in the series are *Astronomy*, *Physical Oceanography*, and *Geology*. Each volume contains a collection of hands-on activities and a series of readings related to the topic area developed for middle/junior high school students.

Overview of Project Earth Science

Project Earth Science was a teacher enhancement program initially funded by the National Science Foundation. Originally conceived as a program in leadership development, this project was designed to prepare middle school science teachers to lead workshops on topics in Earth science. Workshops were designed to help teachers convey key Earth science concepts and content through the use of hands-on activities. With the help of content area experts, concept outlines were developed for specific topic areas, and activities were designed to illustrate those concepts. Several activities were drawn from existing sources; the remainder were developed by Project Earth Science leaders and project staff. Over the course of this two-year project, activities were field tested both in teacher workshops and in classrooms. Participant evaluation played a major role in improving the activities. When completed, activities were reviewed by content experts and organized into a standard format. The curriculum development phase of Project Earth Science received generous funding by BP America, Inc.

About *Project Earth Science: Meteorology*

This book is divided into three sections: activities, readings, and appendix. The activities are constructed around three basic concept divisions. First, students investigate the origin and composition of Earth's atmosphere. Students learn that there is much more to air than meets the eye. Second, students examine some of the factors that contribute to weather. Third, students are introduced to concepts of air masses and the ways these masses interact to produce the weather around us.

A set of readings follows the activities. Some readings are intended to enhance teacher preparation—or serve as additional resources for students interested in further study—by elaborating concepts presented in the activities. Other readings introduce supplemental topics—e.g. the causes of environmental problems such as smog and the greenhouse effect—so that teachers can connect science to broader social issues. Included as an appendix is an annotated bibliography that serves as a supplemental materials guide. Entries are subdivided into various categories: curriculum projects, books, audiovisual materials, instructional aids, and reference materials. Although far from exhaustive, this compilation offers a wide range of instructional materials for all grade levels.

Getting Ready For Classroom Instruction

The activities in this volume are designed to be hands-on and can be performed using materials that either are readily available in the classroom or are inexpensive to purchase. Specific time requirements are noted in each teacher's guide. In general, however, each activity takes one class period or less to complete.

Each activity includes a student unit—ready for duplication—and a teacher's guide. The student section contains background information that introduces the concept behind the activity. It also includes a set of questions to assess student comprehension. These questions are designed to aid understanding, with an emphasis placed on having students draw conclusions about what they have experienced.

The teacher section contains a more detailed version of the background information and a summary of the important points students should understand after completing the activity. In the "Preparation" section, set up for the activity, sources of materials, and alternative materials are described. In "Suggestions for Further Study" the teacher is given ideas for extending studies of topics addressed by the activity, and "Suggestions for Interdisciplinary Reading and Study" includes ideas for relating the concepts in the activity to other disciplines, such as language arts, and social studies. The final section of the teacher's guide provides answers to the questions asked in the student section.

Key Concepts

Activities are organized around three key concepts: the origin and composition of Earth's atmosphere, factors that contribute to weather, and the interaction of air masses. The presentation of concepts and participation in activities should be an integrated process. To facilitate this coordination, a conceptual outline for *Project Earth Science: Meteorology* is presented below.

I. Earth is surrounded by a relatively thin envelope of gases known as the atmosphere. The atmosphere originated billions of years ago and has evolved to its present state.

A. The depth of Earth's atmosphere is relatively insignificant when compared with the radius of Earth.

Activities:

Weather Watch

The Pressure's On

The Percentage of Oxygen in the Atmosphere

Reading:

Earth's Atmosphere

B. The atmosphere originated as a result of outgassing, and its present composition makes life possible.

Activities:

Making Gas

The Percentage of Oxygen in the Atmosphere

It's in the Air

C. Human activity has profound effects on the atmosphere, which in turn affect our climate.

Readings:

The Facts About the Ozone

Air Pollution and Environmental Equity

Global Warning and the Greenhouse Effect

Environmental Effects of Acid Rain

II. Weather is the short term variation of the atmosphere. Some variables that are important in describing weather are temperature, pressure, air movement, and water content in the atmosphere.

A. The sun is the source of energy that drives weather systems within the atmosphere. The amount of sunlight available to heat the ground varies from place to place, hour by hour, and season to season. Differences in available sunlight result in different surface and air temperatures.

Activities:

Why is it Hotter at the Equator than at the Poles?

Which Gets Hotter: Light or Dark Surfaces?

Up, Up, and Away!

B. Differences in air temperature drive air circulation and wind. Global wind patterns are affected by Earth's rotation.

Activity:

Why Winds Whirl Worldwide

Reading:

Weather and the Redistribution of Thermal Energy

C. Water, critical for life, is found in all three chemical phases on Earth: solid (ice), liquid, and gas (water vapor). Although the total amount of water on the planet is constant, it cycles through Earth and its atmosphere, changing phase many times. Large amounts of heat associated with changes of phase help drive cloud formation and storm circulation.

Activities:

Recycled Water: The Hydrologic Cycle

Rainy Day Tales

A Cloud in a Jar

Just Dew It!

Let's Make Frost

It's All Relative

Hail in a Test Tube

Readings:

Weather and the Redistribution of Thermal Energy

Weather's Central Actor: Water

III. A large volume of the atmosphere that has the same temperature and pressure is called an air mass. Interactions between air masses cause many of the weather patterns that we experience.

Activities:

Weather Watch

Moving Masses

Interpreting Weather Maps

Chasing Hurricane Andrew

Readings:

The Inner Workings of Severe Weather

Flash to Bang

Project Earth Science: Meteorology and the *National Science Education Standards*

Effective science teaching within the middle-level age cluster integrates the two broadest groupings of scientific activity identified by the *National Science Education Standards*: (1) developing skills and abilities necessary to perform scientific inquiry, and (2) developing an understanding of the implications and applications of scientific inquiry. Within the context of these two broad groupings, the *Standards* identify specific categories of classroom activity that will encourage and enable students to integrate skills and abilities with understanding.

To facilitate this integration, an organizational matrix for *Project Earth Science: Meteorology* appears on pages 226–227. The categories listed along the X-axis of the matrix, also listed below, correspond to the categories of performing and understanding scientific activity identified as appropriate by the *Standards*.

Subject Matter and Content. Specifies the topic covered by an activity.

Scientific Inquiry. Identifies the "processes of science" (ie. scientific reasoning, critical thinking, conducting investigations, formulating hypotheses) employed by an activity.

Unifying Concepts and Processes. Links an activity's specific subject topic with "the big picture" of scientific ideas (ie. how data collection techniques inform interpretation and analysis).

Technology. Establishes a connection between the natural and designed worlds.

Personal/Social Perspectives. Locates the specific meteorology topic covered by an activity within a framework that relates directly to students' lives.

Historical Context. Portrays scientific endeavor as an ongoing human enterprise by linking an activity's topic with the evolution of its underlying principle.

By integrating the presentation of specific science subject matter with the encouragement of students to organize and locate that subject matter within an accessible framework, *Project Earth Science: Meteorology* hopes to address the *Standards*' call for making science—in this case meteorology—something students do, not something that is done to students. The organizational matrix provides a tool to assist teachers in realizing this goal.

Weather Signs

Evening red and morning grey,
Two good signs for one fine day.
Evening grey and morning red,
Send the shepherd wet to bed.

Dew in the night,
Next day will be bright.

Grey mists at dawn,
The day will be warm.

Rain before seven,
Fine before eleven.

Robin Page

Weather Watch

Background

Without looking out the window, do you know what the weather is like right now? Do you remember what it was when you woke up this morning? Were there any clouds in the sky this morning? If so, what did they look like? If you have trouble answering these questions, you are not alone. Most people do not pay very much attention to the weather until it interferes with something they plan to do. Rain can cancel a baseball game. Snow can cancel school. When these things happen we notice the weather, but much of the time we ignore it. By not paying more attention, we miss many of the interesting things that go on in the atmosphere. For example, it is possible to predict short term weather changes with some accuracy just by looking at the clouds.

The purpose of this activity is to help you observe the weather around you more carefully and to help you relate the weather you're experiencing to weather in other parts of the nation. One way to predict weather changes is to look at the weather in nearby places. You can become an excellent forecaster by carefully observing what is happening around you.

Objective

There are two objectives for this activity. One is to learn weather watching by carefully observing local and national weather for several days. A second is to understand the connection between local and national weather patterns.

Materials

For each student:

- ◊ national weather map every day for two weeks (the teacher will display these)
- ◊ pencil
- ◊ 10 sets of data sheets (one Weather Watch Data Sheet and one map of the U.S. per day) per student. If weekend observations are asked for, more data sheets will be needed.

Procedure

1. Obtain two data sheets from your teacher each day. One has the heading "Weather Watch Data Sheet." The other is a map template of the United States, which you will use as your weather map.
2. Copy the information from the national weather map onto your Weather Map Data Sheet. (The key on the Weather Watch Data Sheet will help you understand the symbols on the national weather map.) Include the temperatures in major cities across the nation.
3. Record the daily weather conditions at your location on the Weather Watch Data Sheet. This includes:

   a. cloud type—refer to the descriptions of clouds on the data sheet. If more than one cloud type occurs in a day, record each and the general times they occurred.

   b. precipitation—record the type (rain, snow, hail, etc.), amount, and duration (steady or intermittent) of any precipitation that occurred during the day. Also record the general times they occurred. If there is no rain gauge at your school, you will have to get some of this information from the weather report on television, radio, or in the next day's newspaper.

c. temperature—record the high and low temperatures for the day from the thermometer at your school. If for some reason you can't use the thermometer at your school, record the daily high and low temperatures from television, radio, or newspaper weather reports. Record this temperature in both degrees Celsius and degrees Fahrenheit.

d. pressure—record the barometric pressure using any available source, such as television weather reports, the National Weather Service's telephone service, or a school or home barometer.

4. Note any unusual weather conditions across the nation or in your location on your Weather Watch Data Sheet.
5. Repeat steps 1–4 for two weeks using a new set of data sheets for each day. (If your teacher wants you to make observations over the weekend, be sure to take home extra copies of the data sheets.)
6. At the end of two weeks, arrange your data sheets chronologically, with the oldest first and the newest last.

Questions/Conclusions

1. Find one of your Weather Watch Data Sheets showing precipitation at your location.

   a. What kind of clouds did you observe on that day?

   b. Check the atmospheric pressure in the days leading up to that precipitation. Did the atmospheric pressure rise or fall? By how much?

2. For the same day that you recorded precipitation, look at the national weather map.

   a. Where in the country were the major fronts (for example, northeast, west, etc.) and what types of fronts were they?

   b. Was there a front near you? What kind of front was it?

3. Are there other Weather Watch Data Sheets showing local precipitation? If so, were the cloud patterns and front movements the same each time it rained/snowed? (Hint: It will be helpful to answer questions 1–2 again for the data

sheet you are using now.) What is the connection between front movements and the precipitation that comes to your area?

4. Look at your first national weather map. Can you find a front or large area of precipitation in the western part of the nation? If not, look through the rest of the weather maps until you find one.

   a. Now look at the national weather maps following the one with the front or precipitation. Has the front or precipitation moved?

   b. If so, in which direction did it move?

   c. Find other fronts or patterns of precipitation and see if they moved in the same direction. Did they? If not, describe the direction in which they moved.

5. From your answers to question 4, would you say there is a general direction of weather movement over North America? If so, what is that direction?

6. Look at the last data sheet you completed. Based on your answer to question 5, what is your prediction for tomorrow's weather in your location?

Weather Watch

Data Sheet

Date:

Cloud Key

Cumulus – individual puffs or heaps with flat bases

Cumulonimbus – large, fluffy, anvil-shaped clouds with thunderstorms

Cirrus – thin, wispy streaks or patches

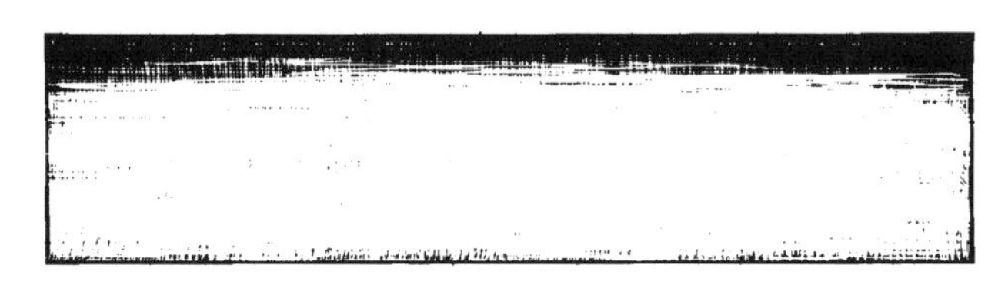

Stratus – uniform gray layer

Weather Map Key (Check your paper's key—symbols may vary slightly)

Warm Front

Cold Front

Stationary Front

High Pressure

Low Pressure

Cloud Observation (Write down appropriate cloud type for each time of day)	Precipitation	Temperature	Pressure
Morning: ____________	**Type:** ____________	**High:** ____________ (°F)	____________ (in. Hg)
Afternoon: ____________	**Duration:** ____________	____________ (°C)	____________ (mm Hg)
Evening: ____________	**Amount:** ____________	**Low:** ____________ (°F)	
		____________ (°C)	

Additional Notes: ____

____

____

____

____

Weather Map

Date:

Weather Watch

Materials

For each student:

- ◊ national weather map every day for two weeks (the teacher will display these)
- ◊ pencil
- ◊ 10 sets of data sheets (one Weather Watch Data Sheet and one map of the U.S. per day) per student. If weekend observations are asked for, more data sheets will be needed.

What is Happening?

People usually pay little attention to the weather. We rely on the media to tell us what weather to expect in the near future. With modern technology, we can access information about the weather with very little effort. This has not always been the case. Long before there were weather satellites, radar, and television, people predicted the weather relatively accurately by carefully observing atmospheric conditions around them over a long period of time.

This activity will encourage students to be more aware of local and national weather. They will also learn that by being very observant, they can predict weather changes with considerable accuracy.

Important Points for Students to Understand

- ◊ Through careful observation of local weather conditions short-term weather changes can be predicted.
- ◊ Weather conditions in other parts of the nation can be used to predict local weather (except perhaps for Alaska, Hawaii, and the far western states).
- ◊ The general pattern of atmospheric movement in the United States is from west to east.

Time Management

This activity will take about ten minutes each day for a period of two weeks. Ten minutes should be enough time for students to record all the required information on their data sheets. At least one class period should be set aside to introduce the activity and another to analyze and discuss the data at the end.

Preparation

Before the activity begins, discuss the movement of weather fronts with the students and list their ideas about weather prediction. Use the list to tailor the activity and support accurate assumptions while correcting misconceptions.

Make arrangements in advance to have national weather maps on hand for a period of two weeks. (These maps appear in all major newspapers.) These maps should include representations of the major weather fronts across the nation. Rather than giving each student a copy of the weather map, make a transparency and use it with an overhead projector. If a newspaper weather map that shows the fronts is not available, it may be possible to videotape the television weather report and show it to the class.

Each student will need two weeks (10 days) of data sheets. (If weekend data gathering is desired, additional sets of data sheets should to be distributed. *The activity will be much more meaningful if weekend data collection is done.*)

The most difficult aspect of this activity for the teacher is finding a two-week period when there are certain to be noticeable weather changes. The National Weather Service gives extended forecasts that include general weather predictions as much as seven days in advance. By noting these, it is easier to find a time when the students will have a chance to see significant weather changes. Late fall through early spring is best in terms of the number of active front systems.

Suggestions for Further Study

Students might take the information on a weather map for granted with no thought of the source of the weather data. Encourage them to find out where the information comes from. A field trip to the closest National Weather Service office (generally at airports or universities), a nearby television station, or a university meteorology department would allow them to talk with meteorologists. Local meteorologists might be willing to visit your class. An interesting student project would be to research how weather maps are constructed. They contain hundreds of pieces of information collected from all over the nation. Activity 17, "Interpreting Weather Maps," has more information on this topic.

Meteorologists in the eastern parts of the United States can predict weather by watching weather systems come across the continent. In other areas, meteorologists do not have this advantage. Encourage students to find out where forecasters in the western part of the continent look for approaching weather.

Only four cloud types are presented to the students in this activity, but there are many others. The activity provides an excellent opportunity for students to learn the classification system for clouds. See the Bibliography for more information on cloud resources.

Answers to Questions for Students

NOTE: Answers to questions 1–3 will vary with students and will also depend on whether they refer to a shower/thunderstorm (a relatively brief period of precipitation) or steady rain/snow. Be sure that students understand the difference between the two. A set of possible answers will be given for each alternative.

1. a. Steady rain/snow: Possibly cirrus and then certainly stratus clouds.

   Showers/thundershowers: Possibly clear skies and then some cumulus and cumulonimbus, in that order.

   Answers, however, can be more complex. In some places (parts of Texas and California, for example), stratus clouds arrive before cirrus, so the stratus clouds can block the view of "expected" cloud sequences. With thunderstorms, there are times when the cirrus "anvil" moves in first and the clouds get lower and darker just before the thunderstorm arrives.

   b. Steady rain/snow: The atmospheric pressure probably dropped as the rain/snow approached.

   Showers/thundershowers: The atmospheric pressure probably dropped as the thunderstorm approached and may have risen rapidly after it passed.

2. a. Answers will vary with students.

   b. Steady rain/snow: There is probably a warm front close by.

   Showers/thundershowers: There is probably a cold front close by.

   Answers, however, can be more complex. Stationary fronts can also give steady rain/snow. Over the western mountains, thunderstorms often form even without the presence of a front.

3. Answers will vary with students. This is an opportunity for both teachers and students to look at the variety of weather patterns and cloud types that may or may not deliver precipitation.

NOTE: Answers to questions 4 and 5 will vary with students and geographic areas. The answers given here are the most likely ones. Other correct answers may be given by students.

4. a. Yes.
   b. To the east or northeast.
   c. Yes.
5. Yes. To the east or northeast.
6. Students should predict that they will experience the weather that is to the *west* of their location.

First Rainfall

On dry days, I remember
the first rainfall on Earth.
Clean and undesigned,
my atoms were there.

Clutching at hydrogen, bloated
with ammonia clouds, the air
no longer held,
and gushed
its first relief.

The wind was wet in sheets,
each force moved in its own sound.
What sod there was turned moist.

The earth, still rounding
like a newborn's head,
gurgled in the fog
and freely outgassed.

Alan P. Lightman

Making Gas

What Is Happening?

Earth's atmosphere today barely resembles the original one. The keys to understanding the evolution of Earth's atmosphere are found in geologic and biological activities. Current scientific evidence suggests that as Earth formed (from accumulating gas, dust, and larger particles), it grew warmer, finally becoming a sphere with a molten core and an atmosphere consisting primarily of hydrogen and helium. The heat trapped in Earth was, and still is, released through geologic activity such as volcanic eruptions. Before a volcano erupts, gases are dissolved in its magma. As the magma comes closer to the surface, the pressure on it decreases. As the pressure decreases, the gases can no longer remain dissolved and are released into the atmosphere (a process called outgassing). Because of Earth's gravity, only the lightest gases (hydrogen and helium) escape to outer space. (Today, the release of gases is monitored at various volcanic sites.)

On the early Earth, as now, the gases released by volcanic eruptions were primarily water vapor, carbon dioxide, and nitrogen. Free oxygen was nearly absent in the earliest atmosphere. Through the combined effects of solar radiation and photosynthesis, the concentration of atmospheric oxygen sharply increased. Intense solar radiation breaks down water into its constituent elements: hydrogen and oxygen. As photosynthetic algae and plants evolved, oxygen was released as a by-product of their metabolism. The intricate combination of these processes on Earth is unique in our Solar System.

Important Points for Students to Understand

◊ Earth's atmosphere has evolved over time and continues to evolve today.

◊ Earth's atmosphere was outgassed from its interior.

◊ Gas dissolved in magma within Earth is released to the atmosphere when the magma rises to the surface due to decreases in pressure.

Objective

The objective is to demonstrate the concept of outgassing and to explain the origins of Earth's atmosphere.

Materials for a demonstration

◊ two effervescent antacid tablets

◊ air tight jar—about 0.5 liter (a canning jar works well)

◊ cold water

◊ paper towels or other absorbent material for spills

Time Management

The water used for this demonstration should be cold initially. This may require refrigerating the water for some time before the antacid tablets are added to it. The antacid tablets should be added to the cold water at least two hours before the demonstration.

Preparation

Before the demonstration, ask students how Earth's atmosphere was formed, and where its constituent gases came from. Inquire if Earth's atmosphere can change or if it has always been as it is today. Questions along these lines can help the teacher tailor this demonstration to address misconceptions and confirm accurate ideas (such as the dynamic nature of the atmosphere).

The day before the demonstration, fill the jar(s) three-fourths full with cold water—the colder the better. Drop the two antacid tablets in the water and immediately seal the jar. Allow the water to warm to room temperature overnight. If enough jars and antacid tablets are available, one jar can be prepared for each student or group of students rather than only for the teacher. The instructions below are written for a teacher demonstration but can be modified to allow the students to participate. Be sure to try this demonstration on your own before doing it for the class. The results can be messy!

Instructions for Demonstration

1. After gaining the students' attention, ask them to describe the contents of the jar and the conditions within the jar. Bring the jar close to them so that they can observe it carefully.
2. Have students write down a prediction for what they think will happen when you open the jar.

3. Open the jar and have the students write down what they observe. Caution: Opening the jar can produce the same results as opening a warm can of soda pop. *Teacher and students should be wearing safety goggles*. Hold the jar away from your clothes. If the activity is being done at each student's desk, there should be something underneath the jar to catch or absorb any water that may spill.
4. Ask students what conditions within the jar changed when the lid was released.
5. Continue with a discussion of how this demonstration models the outgassing of the atmosphere.

Atmosphere

Inscription for a garden wall

Winds blow the open grassy places bleak;
But where this old wall burns a sunny cheek,
They eddy over it too toppling weak
To blow the earth or anything self-clear;
Moisture and color and odor thicken here.
The hours of daylight gather atmosphere.

Robert Frost

The Pressure's On

Background

Air has weight, yet we don't feel it. The weight of the air on Earth's surface produces air pressure. Because we have lived our whole lives exposed to the weight of the atmosphere, we tend to be unaware of its effects. This activity will give you an opportunity to see that air pressure, caused by the weight of the atmosphere, can produce some unexpected results.

Objective

The objective of this activity is to investigate the effects of air pressure.

Materials

Each group will need:

◊ sturdy paper cup

◊ index card

◊ straight pin

◊ water

◊ sink or catch basin

Procedure

Trial 1

1. Working over a sink or a catch basin, fill a cup to the rim with water. In the box marked "Trial 1 Prediction," suggest what will happen when you turn the cup over. Explain your prediction.
2. Turn the cup over. What happened?

Trial 2

1. Fill the cup again. Cover it with the index card, and make sure that you have created a water seal around the rim of the cup, so no air can seep in. In the box marked "Trial 2 Prediction," suggest what will happen when you turn the cup over with the index card covering it. Explain your prediction.
2. While holding the index card on top of the cup, carefully turn the cup over. Hold the cup around the rim at the bottom so that the cup is not deformed (bent) and remove the hand holding the card. What happened?

Trial 3

1. Slowly, turn the cup sideways, holding the edge of the card to keep it in place. Record your observations in the appropriate "Observation" box.

Trial 4

1. Slowly turn the cup so that it is again upside down. Using the straight pin, carefully make a hole in the bottom of the cup and remove the pin. Record your observations.
2. Repeat the Trial 2 procedure, holding a finger over the hole in the bottom of the cup. Can you replicate the results from Trial 2?

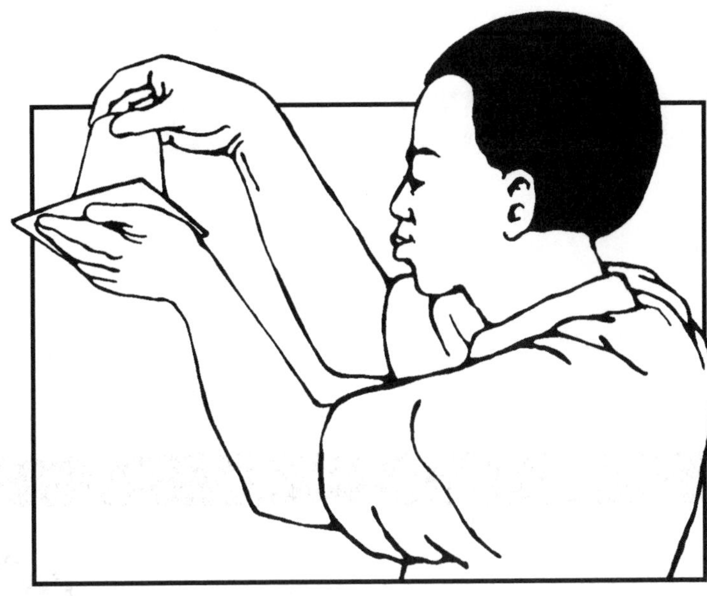

TRIAL 1 PREDICTION

TRIAL 1 OBSERVATION

TRIAL 2 PREDICTION

TRIAL 2 OBSERVATION

TRIAL 3 OBSERVATION

TRIAL 4 OBSERVATION

Questions/Conclusions

1. In Trial 1, what caused the water to fall out of the cup?
2. In Trials 2 and 3, what held the index card to the cup? What prevented the water from falling out of the cup, as it had done in Trial 1?
3. Explain why the water and the index card fell from the cup in Trial 4 of the activity.
4. Based on your observations, in which direction(s) is air pressure being exerted? Draw a picture representing your explanation and explain the phenomenon of air pressure in your own words.
5. Try to explain why we usually do not feel the pressure of the atmosphere around us. When *do* we feel air pressure?

The Pressure's On

Materials

For each group of students:

◊ sturdy paper cup

◊ index card

◊ straight pin

◊ water

◊ sink or catch basin

What Is Happening?

Air pressure is a difficult concept for students to understand. We usually don't *feel* air pressure because we live constantly exposed to it, and our body's structure counteracts its effects. Only when there are rapid changes in pressure—when the plane we are in changes altitude rapidly or when we drive quickly down a mountain road—do we feel the pressure differences. Because they are adapted to much higher pressures, life forms brought up from the bottom of the ocean need to be kept in pressurized containers, or else their bodies literally can explode. Similarly, in the vacuum of space, astronauts need to wear pressurized suits. The activity presented here gives students the opportunity to study the effects of air pressure, and specifically, to experience the fact that air pressure is exerted in all directions.

Pressure is measured in terms of weight (or force) per unit area. In the simpliest case, the air pressure exerted upon a surface—say a tabletop—is equal to the weight of the column of air over that surface extending to the top of the atmosphere. Pressure is not the same as weight, however, because pressure is exerted on a body from all directions, not simply from overhead (just as water pressure is exerted on a fish from all sides). This is due to the fact that molecules in air move randomly and thus are equally likely to travel in all directions.

Air pressure at Earth's surface averages 1.04 kilograms per square centimeter (kg/cm^2). In this activity, students are asked to place an index card over a paper cup and invert the cup. Amazingly (with a proper seal), the water does not fall out of the cup. The *weight* of the water in the cup (the gravitational force exerted on the water) is much less than the force exerted upward by air *pressure* (1.04 kg/cm^2). As a result, the card is held in place. Carefully turning the cup sideways (keeping the seal) has no effect; remember, pressure is exerted from all sides equally. The force exerted by air pressure still is greater than the weight of the water in the cup (see Figure 1).

Figure 1

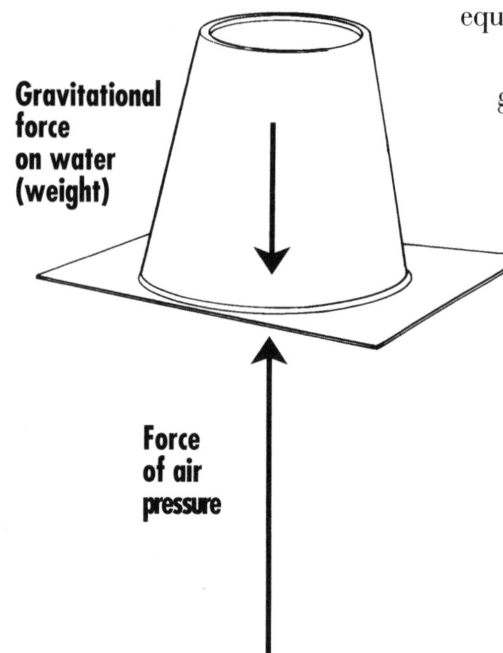

However, when students make a hole in the bottom of the inverted cup, the water falls out of the cup freely because the hole allows air pressure to act on the contents of the cup. Consequently, air pressure pushes down on the surface of the water, neutralizing the upward force of air pressure from below. Thus, the weight of the water in the cup creates a net downward force. The difference between the two forces is the gravitational pull downward being exerted on the mass of water. Hence, the water falls out of the cup.

Give students the time to experiment with this phenomenon, trying a number of variations on the procedure to investigate the effects of air pressure.

Important Points for Students to Understand

- ◊ Air has weight and exerts pressure on everything with which it comes in contact.
- ◊ The force exerted on a surface by air is equal to the weight of a column of air above the surface extending to the top of the atmosphere.
- ◊ Air pressure is exerted equally in all directions.

Time Management

This activity requires less than 15 minutes to complete; however, students should be given the opportunity to fully investigate this phenomenon. This activity works well as a station with other simple activities dealing with air pressure.

Preparation

No special preparations are required for this activity. Teachers may need to assist students who have trouble getting the activity to work. It is important that students work carefully and slowly. A break in the seal between the cup and card allows air into the cup, causing the water to fall. (Note: If students have trouble getting a seal between the cup and the index card, have them fill the cup completely with water and moisten the card slightly before to placing it on the cup.)

Suggestions for Further Study

Have students calculate the weight of the water in the cup and the air pressure on the index card. (Hint: Students will have to calculate the area of the cup's opening.) Challenge them to calculate the maximum height of a water column that could be held up by the index card. (Regardless of the size of the cup, air pressure will support a column of water 10m high.) Also, encourage students to experiment, making the pinhole in other parts of the cup and comparing results.

Have students investigate the limitations of early water wells and how technological advances have been used to overcome these limitations. As mentioned above, air pressure can support a column of water only about 10m high. Early water wells depended only on air pressure to force water from below ground level to the surface. As such, the wells could be no more than about 10m deep. Modern pumps have made it possible to use water sources that are much farther below the surface of Earth.

If students do not realize it on their own, point out to them that 1.04 kg/cm^2 is applied to them as well as to the card and cup. Challenge them to find out how their bodies can withstand such pressure without being crushed. Ask them what they think would happen to their bodies if they were in a place with no atmosphere, e.g. outer space. All students are probably familiar with the phenomenon of "ear popping" that may occur when there is a difference between air pressure and pressure in the middle ear. This is common when taking off and landing in an airplane. An investigation of the processes involved in this phenomenon would make an interesting research project.

Answers to Questions for Students

1. The gravitational pull of Earth forced the water from the cup.
2. Air pressure held the index card to the cup preventing the water from falling out of the cup.
3. A hole in the top of the cup allowed air pressure to be exerted on the top of the water, and this, along with the weight of the water itself, was greater than the air pressure on the bottom of the index card alone.
4. Air pressure is equal in all directions.
5. Answers will vary; however, the important point for students to have is that their bodies are equilibrated to their surrounding pressure and adjusts to changes in pressure. Sometimes pressure changes occur too rapidly and the adjustment lags behind. This is when we feel the atmosphere's pressure.

From *This Island Earth*

"Instrumental analysis would show that Earth's atmosphere is a combination of nitrogen and oxygen, lightly spiced with carbon dioxide, argon, and trace elements. It is thick enough to serve as a global heat transfer mechanism and as a shield against lethal energetic particles and bits of rocky solar system debris racing in at hypersonic velocities. But for all its complex and ever-changing white turbulence, the atmosphere would also be frequently semitransparent, giving the planet a window on the universe."

NASA

The Percentage of Oxygen in the Atmosphere

Background

The atmosphere is made up of several different gases. Two of these gases, nitrogen (N_2) and oxygen (O_2), account for about 99 percent of the atmosphere. The rest consists of very small amounts of other gases such as argon (Ar), carbon dioxide (CO_2), and water vapor (H_2O). Each of these gases is important, but we are probably most aware of the importance of oxygen.

Would you predict that oxygen makes up most of the atmosphere? There are several ways to determine the percentage of oxygen in the atmosphere. In this activity, you will take advantage of a chemical reaction—between iron and oxygen that forms rust—to find out just how much oxygen there is in the air.

Objective

The objective of this activity is to determine the percentage of oxygen in the atmosphere.

Materials

For each group of students:

- ◊ 2 test tubes
- ◊ 1 600 ml beaker
- ◊ several grams of iron filings
- ◊ ring stand
- ◊ 2 utility clamps
- ◊ 100 ml graduated cylinder
- ◊ glass-marking pencil

Procedure

1. Predict what percentage of the atmosphere you think is oxygen, and record this in Data Table 1 under the column "Predicted Percentage of Oxygen in the Atmosphere."
2. Fill one of the test tubes with water, and then pour this water into the graduated cylinder. Read the volume and record it in Data Table 1 under the column "Initial Volume of Water in the Test Tube."
3. Take a small quantity of iron filings and sprinkle them in the bottom of the damp test tube. Don't worry if some of the filings get stuck to the side of the test tube.
4. Fill the beaker about two-thirds full of water and place it next to the ring stand.
5. Assemble the test tubes, beaker, and ring stand as shown in Figure 1. The second test tube is a control that will allow you to compare what happens when no iron filings are present.
6. Lower the test tubes into the water until the mouths of the test tubes are just below the level of the water, as shown in Figure 1. Note the water level in the test tubes.

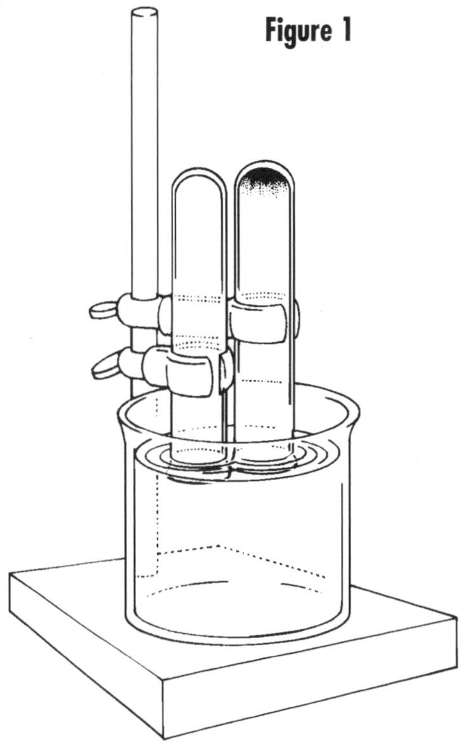

Figure 1

7. Allow the apparatus to sit for 24 hours. After this time, mark the water level in the test tube containing iron filings with the glass-marking pencil without disturbing the test tube. Also observe the appearance of the iron filings and the water level in the control test tube, recording these in Data Table 2 under the column "Observations."
8. Continue marking the water level and recording observations about the iron filings test tube and the control test tube every 24 hours until the water level in the test tube with filings does not change for two consecutive days.
9. Remove the test tube from the beaker and clamp, and rinse the iron filings out of it.
10. Fill the test tube with water to the level of the last line that you marked with the glass-marking pencil. Pour this water into the graduated cylinder. Record this volume in Data Table 1 under the column "Final Volume of Water in the Test Tube."
11. Subtract the final volume of water in the test tube from the initial volume and record this in the column labeled "Volume of Oxygen."
12. Calculate the percentage of oxygen in the atmosphere by using the following equation. Record your answer in Data Table 1.

$$\text{Percentage of oxygen in air} = \frac{\text{Volume of oxygen}}{\text{Initial volume of water in the test tube}} \times 100$$

Questions/Conclusions

1. Explain your prediction for the percentage of oxygen in the atmosphere.
2. What happened to the iron filings in the test tube over the course of the experiment? Why?
3. What happened to the level of water in the test tube over the course of the activity? Why do you think this happened? (Hint: Why does the level of water in a straw rise when you suck on the straw?)

4. How does the percentage of oxygen in the atmosphere you calculated compare with what you predicted it would be? Explain the difference if there is one.
5. According to your calculations, what percentage of the atmosphere is nitrogen? Are you surprised by this result?

DATA TABLE 1

Predicted percentage of oxygen in the atmosphere	Initial volume of water in the test tube	Final volume of water in the test tube	Volume of oxygen	Percentage of oxygen in the atmosphere

DATA TABLE 2

Day	Observations
1	
2	
3	

The Percentage of Oxygen in the Atmosphere

Materials

For each group of students:

- ◊ 2 test tubes
- ◊ 1 600 ml beaker
- ◊ several grams of iron filings
- ◊ ring stand
- ◊ 2 utility clamps
- ◊ 100 ml graduated cylinder
- ◊ glass-marking pencil

What is Happening?

Earth's atmosphere consists of several gases, but nearly 99 percent of it is just two gases: nitrogen (N_2) and oxygen (O_2). Students often think that the atmosphere is all oxygen or mostly oxygen simply because they have heard the most about it, and they know they need to breathe oxygen in order to live. However, only 20.9 percent of the atmosphere is oxygen.

A simple chemical reaction can be used to demonstrate this fact to students. The oxygen in the atmosphere reacts with many things in the environment in a process known as oxidation. The process actually creates a new substance with different properties. We are most familiar with the oxidation of metals; it commonly is called corrosion or rust. Living things are oxidized, too. Apples and lettuce leaves that have been cut turn brown when exposed to the air. This is a result of oxidation, too.

The specific reaction being observed in this activity is the oxidation of iron. Iron, water, and oxygen combine to produce rust (iron hydroxide). Here, iron filings are used to consume all the oxygen in a sample of air. Students can determine the percentage of oxygen in the atmosphere by measuring the volume of the air before and after it reacts with the iron filings.

Important Points for Students to Understand

- ◊ Earth's atmosphere is composed of several gases, not just oxygen.
- ◊ Oxygen comprises about one-fifth of Earth's atmosphere.
- ◊ Oxygen is a chemically active gas.

Time Management

The time required to set up this activity is less than one-half of a class period. Daily observations (about 5 to 10 minutes) are required for several days. One-half of a class period can be used at the end to compute and analyze the results.

Preparation

Be sure to have all the materials required for this activity either centrally located or already distributed to the groups. Since the activity takes more than one class period to complete, it may be difficult to have more than one class of students doing the activity at once. If more than one class does the activity at once, mark all the beakers clearly so that they do not become disorganized. Alternatively, the activity may be done as a demonstration by the teacher for all the classes.

Percentages may be a difficult topic for younger students. Some preliminary class work with percentages will make this activity more meaningful for younger students.

Suggestions for Further Study

The method presented in this activity is but one way to determine the percentage of oxygen in the atmosphere. Students might be interested in trying at least two methods and comparing their results.

It would also be interesting for students to research what the consequences would be if the percentage of oxygen (or other gases) in the atmosphere were to change significantly. Such an investigation may lead them to topics such as the greenhouse effect (Reading 6), depletion of the ozone layer (Reading 4), and acid rain (Reading 7).

Answers to Questions for Students

1. Answers will vary with students.
2. The iron filings became oxidized (or corroded—rust formed on them). The iron reacted with the water and the oxygen in the air.
3. The level of water rose in the test tube. Originally, the air pressure outside the test tube and inside it were the same. However, oxygen was removed from the air in the test tube as it formed rust on the iron filings. As this happened, air pressure in the test tube became lower, and the water was forced up into the test tube by the air pressure outside the test tube.
4. This answer depends on the students' predictions.
5. This answer will depend on their calculation of the percentage of oxygen in the atmosphere. The percentage of oxygen in the air should be subtracted from 99 percent to get the answer they report in this question.

Smog

Who is
making you sick?
Choking you with the stench
of smoke that rises up from the
cities

by day
stifling you with
fumes and vapors that burn
with a brown haze, and poison that
mottles

and plagues
you with cankers,
coughing dust in your face
until your flushing cheeks burn with
fever?

Myra Cohn Livingston

It's In The Air

Background

The atmosphere is made up of several gases, mostly oxygen (O_2) and nitrogen (N_2). However, there are also a lot of particles in the atmosphere. This may at first make you think of pollution. To be sure, much of the particulate matter *is* the result of humans polluting the air, but there are also natural sources of particles. Wind blown dust and pollen are examples. It is necessary to have some particles in the atmosphere because they play an important natural role in the formation of clouds, snow, and rain.

It may be difficult to believe that there are particles in the atmosphere. Most of them are too small to observe just by looking in the air. There is a simple way to make them easier to see.

Objective

The objective of this activity is to investigate the amount and types of particulate matter in the air.

Materials

Each group will need:

- ◊ 2 white coffee filters
- ◊ strainer
- ◊ magnifying glass
- ◊ shallow pot, pan, bucket, or cookie sheet with sides
- ◊ white dinner plate (disposable paper plates work well)
- ◊ tap water

Procedure

1. Using a magnifying glass (or microscope if you have one), observe one of the filters carefully. Describe in the box marked "Unused Filter" what you see.
2. Label one of the coffee filters to identify your group (perhaps with everyone's first initial) and also with a "C" to designate this as the control filter. Place this filter in the strainer, or if you have an automatic drip coffee maker, you can use its holder instead of the strainer.
3. Fill the pot or pan half full with water, and then immediately strain this water through the filter. This trial serves as your control experiment to determine what, if any, particles are present in your water. With this information, you can more accurately determine how many particles come from the air when you get the results of the second part of this experiment.
4. Carefully remove the filter and spread it out on the plate.
5. Allow the plate to sit undisturbed until the filter is completely dry. Don't place the plate near an air vent or open window where a breeze could blow stray particles onto your filter paper or blow any particles on your filter away. (Either would ruin your control.)

6. Now, for the second part of this experiment, fill the pot or pan again with approximately the same amount of water as you used for the control experiment above.
7. Don't strain this water yet. Set this water outside, in a place where it will not be disturbed. Leave the pot in this place for at least one day, two if possible. (If you can't go outside, place your pan near an air vent or open window.)
8. Now, back to the first part of this experiment. When your first filter—your control filter—is completely dry, use a magnifying glass (or microscope, if you have one) to examine the filter carefully. Without the magnifying glass, you may miss many of the particles.
9. Describe in the box marked "Control Filter" what you see.
10. When you are ready to strain the pot of water that has been exposed to air for a day or two, label the second coffee filter for your group and put an "E" on it for "experimental." Place this filter in the strainer.
11. Take the filter and strainer outside to the pot, and carefully pour the water through the filter.
12. After the water was filtered through the strainer, take the filter and pot inside. Rinse the sides of the pot with a small amount of water and pour this through the filter too.
13. Carefully remove the filter and spread it out on the plate to dry. As with the first filter, allow the plate to sit undisturbed until the filter is completely dry. Again, be sure that there is no breeze near the filter while it dries; otherwise, your experiment will be ruined.
14. When the experimental filter is completely dry, use your magnifying glass (or microscope) to examine the filter carefully.
15. Describe in the box marked "Experimental Filter" what you see. Compare what you see on this second filter paper with what you saw on the control filter.

Questions/Conclusions

1. What was your reaction when you examined the second coffee filter (labeled "E")? Were there more particles than you expected? Fewer? What about the types of particles? The colors, the sizes?

2. Where do you think the different kinds of particles came from?
3. Some of the beneficial aspects of particles in the air have already been mentioned. What harmful effects of these particles can you imagine?
4. What time of year did you do this activity? When do you think you would have found more particles and when would you have found fewer? Why?
5. What weather conditions would increase the amount of particles in the atmosphere? Decrease the amount?
6. Are there areas of your city or town where you think air pollution particles would be worse than where you did the activity? If so, where, and why do you think this?

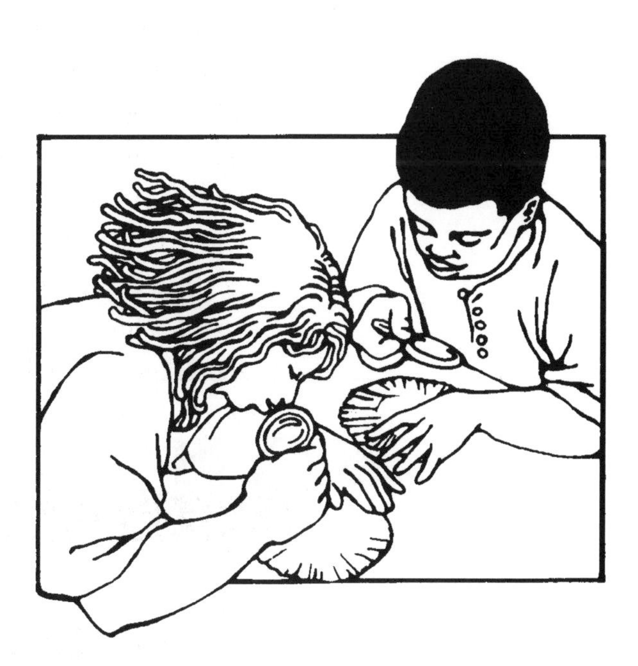

UNUSED FILTER

CONTROL FILTER

EXPERIMENTAL FILTER

It's In the Air

What Is Happening?

In addition to the gases in the atmosphere, there is also a significant amount of particulate matter. Among other sources, these particles come from volcanoes, wind erosion of soil, forest fires and burning in general, and from both industry and agriculture. These particles have both beneficial and harmful effects. They provide the surfaces on which cloud droplets, raindrops, hailstones, and snowflakes form. "A Cloud in a Jar" (Activity 12) and "Hail in a Test Tube" (Activity 18) discuss this aspect of the atmosphere in more detail.

The weather influences the amount of particles in the air. Because cloud droplets form on particles, any type of precipitation will carry particulate matter out of the air to the ground, in a sense cleaning the air. Very windy conditions will do just the opposite, carrying large amounts of particulate matter into the air. This can cause respiratory and visibility problems.

It may be difficult to convince students of the presence of particulate matter in the atmosphere since they cannot readily see it. In this activity, students will use simple filtration to investigate the amount and kinds of particulate matter in the atmosphere.

Materials

Each group will need:

- ◊ 2 white coffee filters
- ◊ strainer
- ◊ magnifying glass
- ◊ shallow pot, pan, bucket, or cookie sheet with sides (something that will offer a large surface area of water)
- ◊ white dinner plate (disposable paper plates work well)
- ◊ tap water

Important Points for Students to Understand

- ◊ Although their small size prevents us from seeing them, significant numbers of particles are present in the atmosphere.
- ◊ The particles can have both beneficial and harmful effects.
- ◊ Airborne particles affect cloud formation, in turn affecting water and weather cycles. Conversely, weather affects the number of particles in the air.

Time Management

This activity will take at least two days, possibly more. However, most of the work can be done by the students at their homes. This activity can easily be done without teacher supervision as long as the instructions are made clear. The materials (with the exception of the magnifying glass) are likely to be found in the students' homes. Furthermore, by having the students do the activity at home, the results from different locations (near a factory, near a highway, in farmland, etc.) can be compared. Students should be required to bring their coffee filters to school. Separately "laminating" them with plastic food wrap is one way to make the filters easily transportable.

Preparation

Little preparation is required for this activity. Most importantly, if the activity is to be done at home, the students must be very clear about the instructions. Some tap water has a significant quantity of particulates in it. If this is encountered, it would be best to use distilled or prefiltered water. Depending on the region of the continent, there may be times of the year when the odds of detecting large quantities of particles are greater than usual. For example, in the southeast, spring is a good time to do the activity because of the pollen in the air.

If a stereoscope (dissecting microscope) is available, it may be even more useful than a microscope in examining particles on the coffee filters. As an option to using coffee filters, the water may just be observed in the pot. Some particles may be even more visible in this way.

Suggestions for Further Study

Repeat the experiment, but in a more quantitative fashion. Draw a grid on the filter paper, then count the particles in some of the squares and calculate the average. Then, by finding the area of the filter paper, students can calculate the number of particles in the water sample. Particle counts such as these are included regularly in weather reports. Students can monitor these in conjunction with this activity.

Students can investigate why some areas have more particulate matter in the air than others and discuss ways to reduce amounts of particulate matter. Encourage students to consider the ways in which they contribute particulate matter to the atmosphere.

Suggestions for Interdisciplinary Reading and Study

Much has been written about air pollution, particulate and otherwise. For a concise overview of one form of air pollution, see Reading 3, "Smog: Its Nature and Effects." Smog also has been the subject of poetry; Myra Cohn Livingston's poem "Smog" may be used to introduce the topic. Reading 4, "Urban Air Quality: The Problem," explains why air pollution continues to be so persistent a problem, despite twenty-five years of control efforts.

The human body's respiratory system has effective filtering mechanisms to prevent some particulate matter from entering the lungs. This activity provides an opportunity to talk about the respiratory system and the effects of particulate matter on the human body.

Answers to Questions for Students

1. This answer depends on the student. They should be encouraged to give very detailed descriptions of the particles.
2. There are no "wrong" answers here. Some are more reasonable than others.
3. Problems in breathing. Allergies. Visibility.
4. Answers will vary with students. Spring and fall are times when a lot of particles are in the air due to pollen, but this depends on the location. In western and southwestern states, winter and summer are better.
5. Windy or dry conditions will increase the amount of particulate matter in the air. Any type of precipitation will decrease the amount.
6. Answers will vary with students. Areas that might have more particulate matter are those near roads, those in heavily populated parts of the city or town, or those downwind from factory smokestacks.

Sun Song

Sun and softness,
Sun and the beaten hardness of the earth,
Sun and the song of all the sun-stars
Gathered together–
Dark ones of Africa
I bring you my songs
To sing on the Georgia roads.

Langston Hughes

Why Is It Hotter at the Equator Than at the Poles?

Background

Two factors are crucial for determining the temperature of Earth's atmosphere and surface. One is day length. In summer, days are longer than nights (i.e. the time between sunrise and sunset is longer), and the total amount of energy received from the sun is high. In winter, the number of daylight hours is reduced, and the total energy received from the sun is decreased. While the change in the number of daylight hours near the equator is relatively small, the change becomes extreme at the poles. Sometimes the sun does not set; other times it does not rise!

But if day length was the only influence on Earth's surface temperature, the poles would be the hottest places on the planet during their summer times. This does not happen. Why?

The other crucial factor affecting Earth's surface temperature involves the angle at which sunlight strikes Earth. Because Earth's axis is tilted to a 23.5 degree angle and because of Earth's shape, sunlight strikes its surface at different angles in different places and in the same place at different times of the year (see Figure 1). This activity is designed to explore how changes in the angle of incoming sunlight affects the way objects are heated. Such differences are a crucial feature of our planet's temperature mechanism.

Objective

The objective of this activity is to investigate the different heating effects of sunlight.

Materials

For each group of students:

- 3 identical Celsius thermometers (glass or metal backed)
- reflector lamp with clamp and 60 watt bulb
- ring stand with iron ring
- utility clamp
- black construction paper (one sheet per student or group of students)
- stapler
- several books to prop thermometers
- meter stick
- scissors

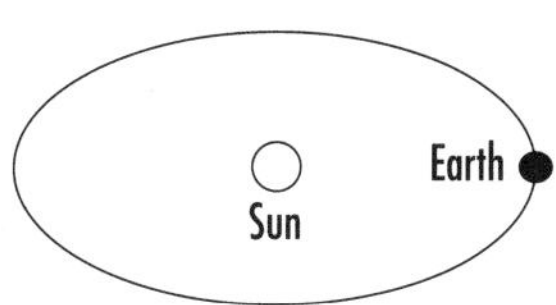

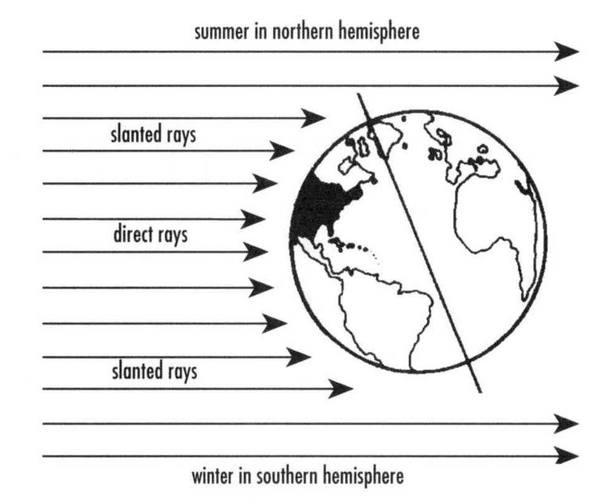

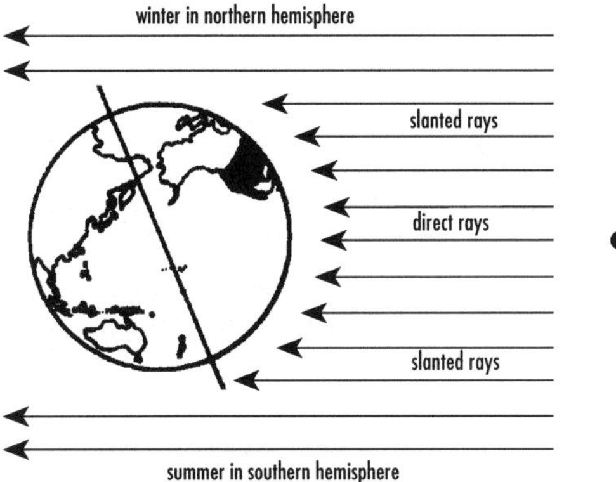

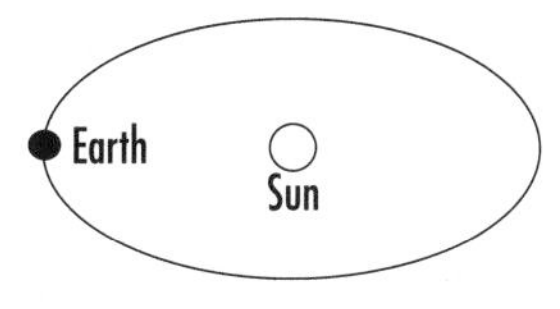

Figure 1. The amount of radiant energy absorbed on Earth depends on the number of daylight hours and on the incoming angle of solar rays. Compare the incoming angle of solar rays during winter and summer in the different hemispheres.

Procedure

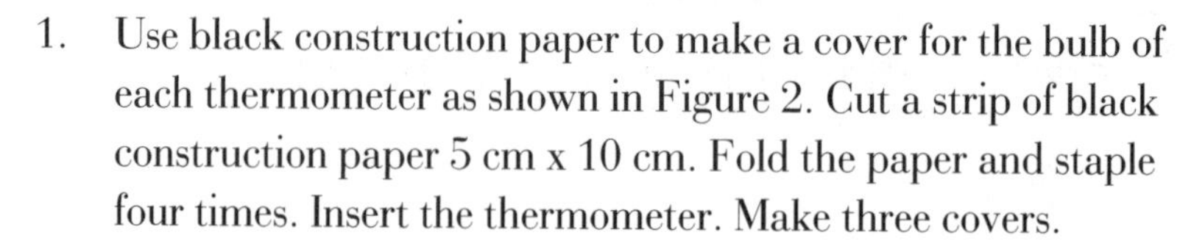

Figure 2

1. Use black construction paper to make a cover for the bulb of each thermometer as shown in Figure 2. Cut a strip of black construction paper 5 cm x 10 cm. Fold the paper and staple four times. Insert the thermometer. Make three covers.
2. Prop the thermometers as shown in Figure 3. One thermometer should be vertical (A), one slanted at about a 45 degree angle (B), and one horizontal (C). Make sure you can easily read the scales without touching them during the experiment.
3. Attach the lamp to a ring stand, being sure it will not move during your experiment. Adjust the lamp on the stand so that its bulb is centered 40 cm above the bulbs of the thermometers.
4. Before turning on the lamp, record the temperature of all three thermometers in the Data Table under the "0 minutes" column.
5. Turn on the lamp and record temperatures for each thermometer every minute for 15 minutes. Do not move the thermometers when reading the temperatures. Record all temperatures in the Data Table.
6. Using the graph paper found at the end of the activity, make a graph of *temperature versus time* for each thermometer. To make comparison easier, plot the results for all three on the same graph paper, with different lines (solid vs. dashed) to show the results from each thermometer.

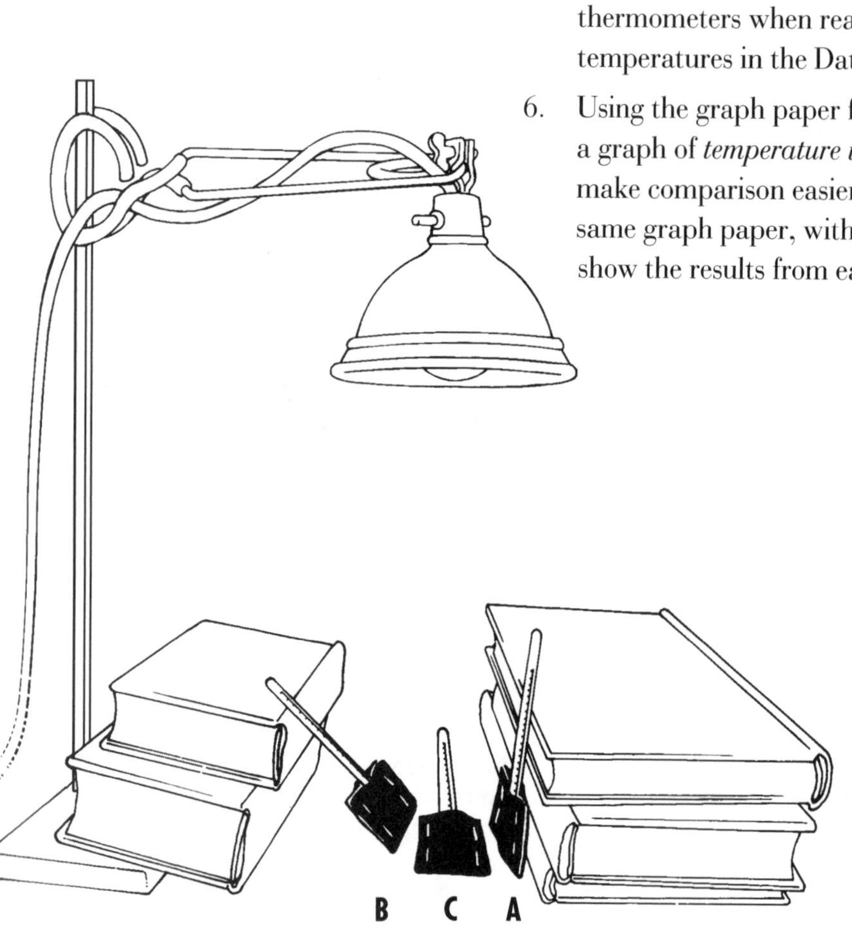

Figure 3

Questions/Conclusions

1. Which thermometer showed the greatest temperature increase? Why?
2. Which thermometer(s) best represents the way sunlight strikes the equator? the poles? What parts of the globe would the third thermometer represent?
3. Using what you learned in this activity and the illustration in Figure 1, how can you explain the fact that the equator is always hotter than the poles?
4. If you were given a data table that listed the average yearly temperatures for cities as you go away from the equator, do you think you would see a trend in the temperatures? If so, what would this trend be and why would it exist?

DATA TABLE

Time (min.)	0	1	2	3	4	5	6	7	8	9	10	11	12	13	14	15	Total change in temperature
Thermometer A																	
Thermometer B																	
Thermometer C																	

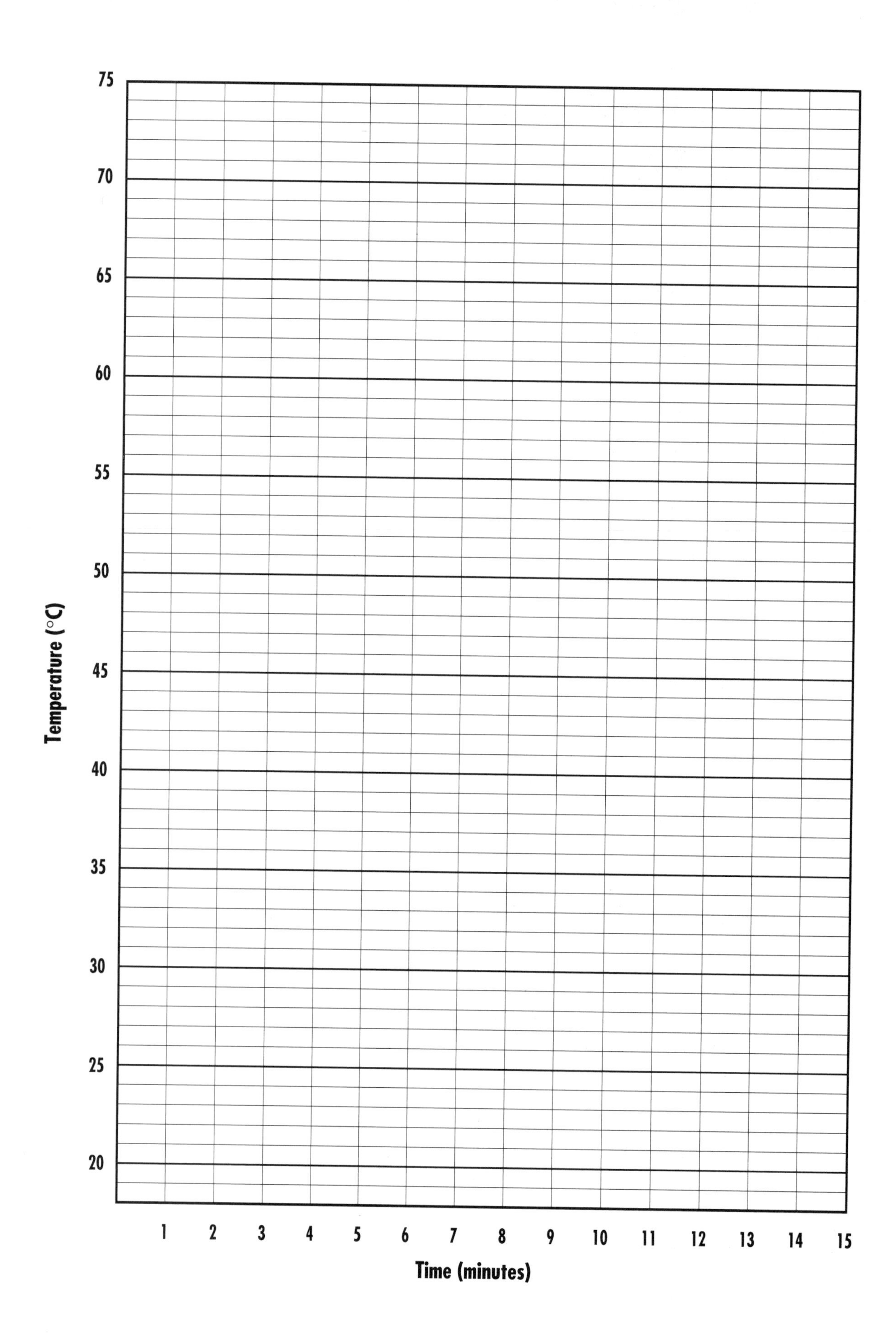

75
70
65
60
55
50
45
40
35
30
25
20
Temperature (°C)
1
2
3
4
5
6
7
8
9
10
11
12
13
14
15
Time (minutes)

Why Is It Hotter at the Equator Than at the Poles?

Materials

For each group of students:

- ◊ 3 identical Celsius thermometers (glass or metal backed)
- ◊ reflector lamp with clamp and 60 watt bulb
- ◊ ring stand with iron ring
- ◊ utility clamp
- ◊ black construction paper (one sheet per student or group of students)
- ◊ stapler
- ◊ several books to prop thermometers
- ◊ meter stick
- ◊ scissors

What Is Happening?

Earth's shape determines how its atmosphere and surface are heated. Equatorial regions receive the maximum amount of solar radiation; polar regions receive the minimum. This experiment demonstrates the principle involved.

When a piece of paper is parallel to a light source (as is thermometer A), the light rays strike the paper at a very low angle, and the radiant energy is dispersed over a relatively large area. When the paper is tilted away from parallel (thermometer B), increasingly direct rays strike the paper's surface and the radiant energy is increasingly concentrated. The more perpendicular the paper is to the incoming light, the more heat it will receive and the quicker it will heat. The paper covering thermometer C will heat quickest.

The dependence of light's concentration on its incoming angle can be demonstrated using a flashlight and a piece of graph paper. Holding the flashlight fairly close (10-15 cm) and perpendicular to the graph paper, trace the outlines of the beam striking the surface. Slowly tilt the paper (trying to keep the flashlight in the same place), occasionally tracing the new area covered by the beam. As the surface area covered by the light increases, the radiant energy from the flashlight beam is dispersed. As a result, the heating effect of the incoming light will be diminished.

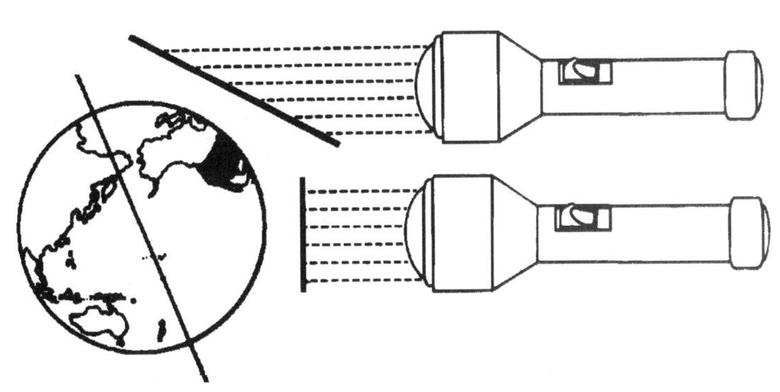

When rays of light strike the surface of an object at a 90 degree angle (i.e. perpendicular), the light can be called "direct." When the angle is anything less than 90 degrees, the light is "slanted" or indirect. While these two definitions are helpful, it is better to think of the light striking Earth along a continuum from most slanted to most direct.

It is important to point out that while the poles receive more slanted sunlight, they may receive it for an extended period of time due to the tilt in Earth's axis of rotation. Because of this, parts of Alaska will actually be warmer at times than some locations much closer to the equator. This activity simply demonstrates that slanted light does not heat objects *as quickly* as direct light.

If Earth was not rotating, this differential heating would set up one global circulation cell with warm air at the equator rising and travelling toward the poles and cold air at the poles sinking and moving toward the equator. Earth's rotation makes for a much more complex global circulation pattern that involves several convection cells (see Reading 5, "Weather and the Redistribution of Thermal Energy").

Important Points for Students to Understand

◊ Slanted light does not heat objects as quickly as direct light.

◊ Because Earth is nearly round, the equator receives direct light, and the poles receive slanted light, with a gradation in between.

◊ Due to the differential heating of Earth's surface, it is always warmer at the equator than at the poles.

Time Management

This activity takes one class period or less to conduct. Preparing the materials takes as much time as actually conducting the activity. Therefore, the activity can be done in less time if materials are prepared in advance.

Preparation

Be sure that all materials are either centrally located or already distributed to student groups. The teacher may do as much or as little preparation of materials (setting up lamps, covering thermometers) as is desired. The more preparation done ahead of time, the less time will be required for the activity, but the need for students to learn to use laboratory equipment should be considered. Use alcohol-filled thermometers for this activity, and *urge students to use caution* to avoid breaking the fragile thermometers and burning themselves on the lamp. Having a globe on hand may aid your discussion.

Suggestions for Further Study

Ask students why black construction paper was used. Also ask them why the same color was used on each thermometer. Why not use different colors? The fact that different colored surfaces heat at different rates is another important factor in weather (see activity 7, "Which Gets Hotter: Light or Dark Surfaces?").

Students can use the newspaper to research average temperatures of locations around the world and construct a graph/map of average yearly temperatures by latitude to investigate any trends. Abnormalities in the trends may lead them to investigate other factors that affect temperature, e.g. elevation and proximity to oceans.

Encourage students to investigate the months-long days and nights at the poles. This may force them to take a more global view of Earth and should lead them to discover the tilt in Earth's axis of rotation and its relationship to seasons. This activity leads naturally into a discussion of seasons and the misconceptions that surround the topic of seasons.

Depending on the level of the students, they may be able to reason that the atmosphere is heated differentially just as the surface of Earth is. It is this differential heating of the atmosphere, together with the rotation of Earth, that sets up the global circulation patterns of the atmosphere. Encourage students to make predictions and hypotheses about global circulation patterns and then do research to find out if their hypotheses are reasonable.

Suggestions for Interdisciplinary Study

Of course, seasons in the northern and southern hemispheres are reversed. This activity may provide a launching point for investigating cultural activities at the present time in the other hemisphere.

Investigating how people live under conditions where the pattern of day and night is radically different from that at your latitude provides another launching point for investigating other cultures. Exploring how other groups adapt to seasonal changes in heating patterns (or to having little seasonal change, as near the equator) offers similar opportunities.

Answers to Questions for Students

1. The horizontal thermometer. Because it was the one receiving the most direct rays.
2. The horizontal thermometer (C) represents the equator; the vertical thermometer (A) represents the poles; the slanted thermometer represents about 45 degrees latitude (approximately the latitude of Tasmania and southern Argentina in the southern hemisphere or South Dakota and Romania in the northern hemisphere).
3. The equator receives the most direct rays of sunlight, and the poles receive the least direct rays. As this experiment shows, more direct light causes quicker heating than less direct light. Therefore, all other things being equal, the equator is always hotter than the poles.
4. Answers will depend on students but they should see a trend. That trend should be one of general decrease in average yearly temperature as latitude increases towards the poles. Moving away from the equator, Earth's surface receives less and less direct sunlight. If students look at actual data, the trend may be less clear in areas close to major geographic features, e.g. large lakes, oceans, or mountains. These geographic elements also affect temperature.

It Bids Pretty Fair

The play seems out for an almost infinite run.
Don't mind a little thing like the actors fighting.
The only thing I worry about is the sun.
We'll be all right if nothing goes wrong with the lighting.

Robert Frost

Which Gets Hotter: Light or Dark Surfaces?

Background

Do you ever walk barefoot on warm, sunny summer days? If so, are there surfaces that you avoid so you won't burn your feet? On days like this, is it better to wear light or dark colored clothing? Which type of clothing is best to wear on cold, sunny winter days if you want to get warm quickly? You probably have answered some of these questions before. In this activity, you will learn the reason behind those answers.

Objective

The objective of this activity is to investigate the rates at which different colors of the same surface heat.

Materials

For each group of students:

◊ reflector lamp with 100 watt bulb

◊ 1 black metal cup or can

◊ 1 white metal cup or can

◊ two insulated lids with slits

◊ two thermometers

◊ ruler or meter stick

Procedure

1. Slide the thermometers through the slits in the insulated lids so that the bulb of each will be about half-way down in the cup, as shown in Figure 1. *Caution: Do not force the thermometers into the slits.* If they will not go in, make the slits slightly larger.
2. Place the lids on the cups.
3. Place the cups side-by-side so that each is about 10 cm from the lamp as shown in Figure 1. Keep the lamp turned off for now.

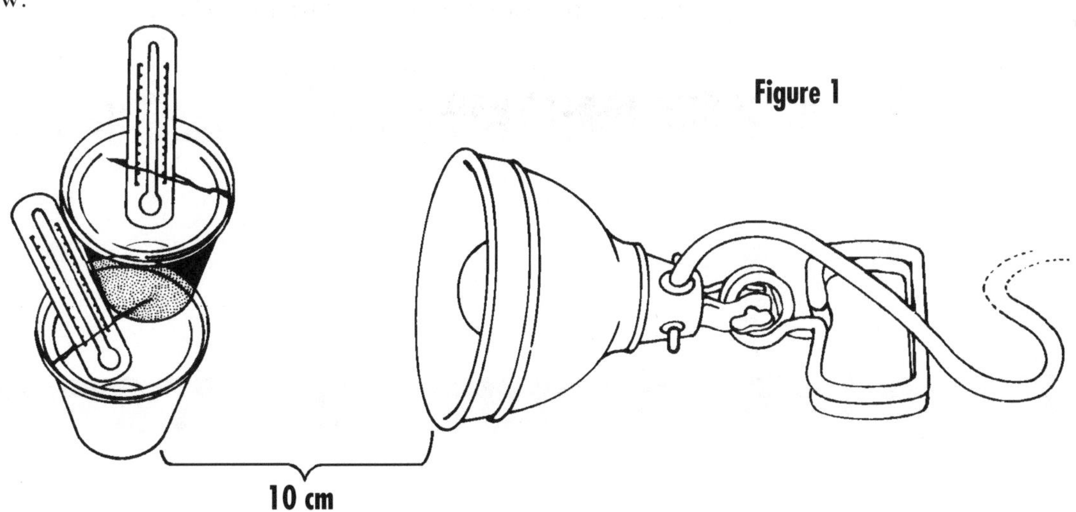

Figure 1

4. Record the initial temperature of each cup in the Data Table under the "0 minutes" column. In the box marked "Prediction" suggest what will happen to the temperature in each cup when the light is turned on, and explain your prediction. Will there be any differences between the cups?

5. Turn on the light and record temperatures every minute for 5 minutes. While the light is on, do not touch, move, or disturb the cups in any way.
6. Graph your results on the graph paper provided.

Questions/Conclusions

1. What did you notice about the temperatures of the two cups? How did this compare with what you predicted in step 4?
2. From your observations, what color surface heats most quickly? Why?
3. On a hot, sunny summer day, which would you rather wear outdoors—a white shirt or a black shirt? (Assume they are made of the same material.) Why?
4. A glider is a kind of airplane, one without an engine. In order to stay in the air, glider pilots sometimes look for large, paved areas (or fields that have been plowed recently) to fly over. Why do they do this?
5. How could a large, light-colored area of land and a nearby, dark-colored area of land create wind on a bright, sunny day?
6. What color roof would you want in a cold climate? In a hot climate? Why?

PREDICTION

DATA TABLE

Time (min.)	0	1	2	3	4	5
Black cup temperature (°C)						
White cup temperature (°C)						

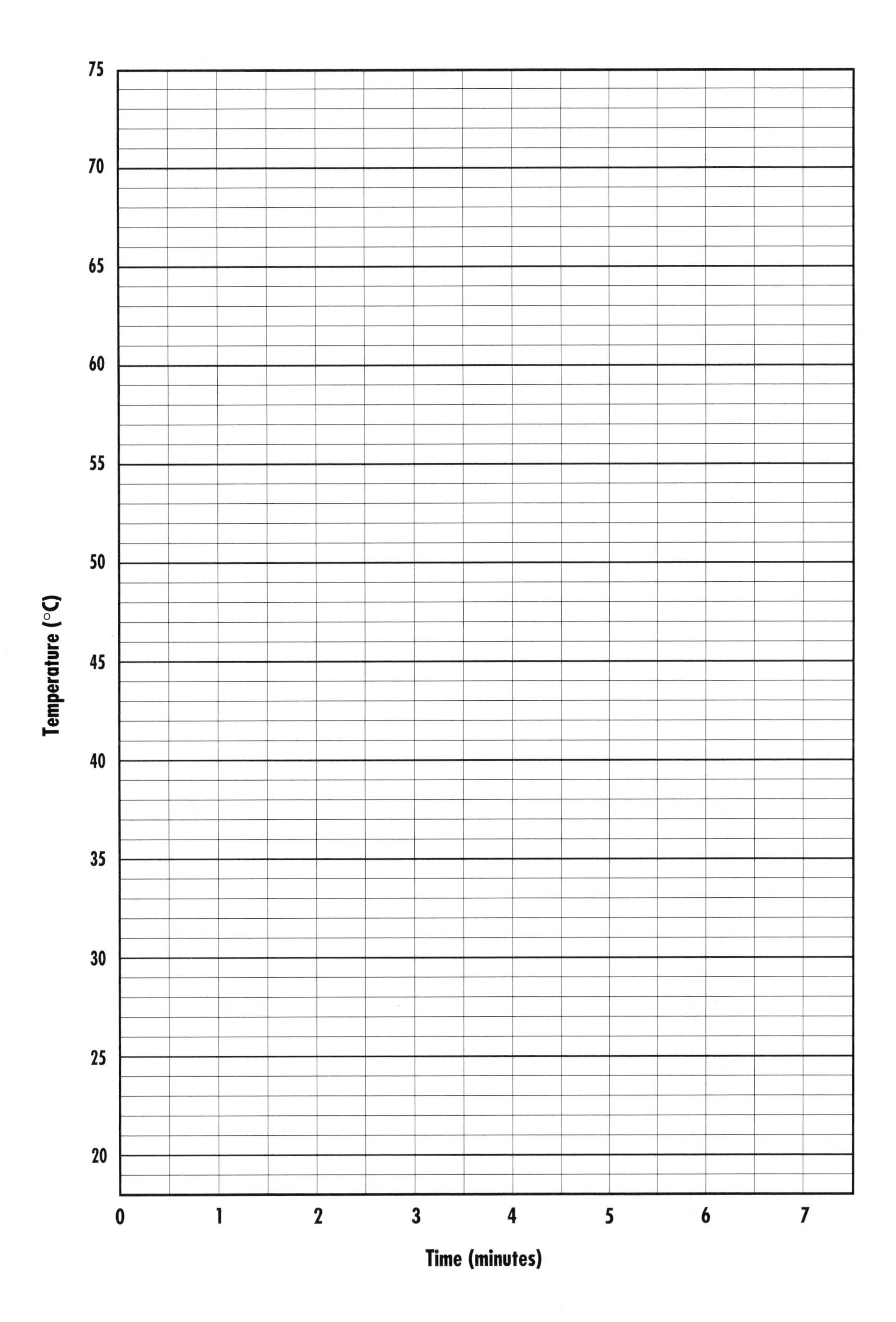

75
70
65
60
55
50
45
40
35
30
25
20
Temperature (°C)
0
1
2
3
4
5
6
7
Time (minutes)

Which Gets Hotter: Light or Dark Surfaces?

Materials

For each group of students:

- ◊ reflector lamp with 100 watt bulb
- ◊ 1 black metal cup or can
- ◊ 1 white metal cup or can
- ◊ two insulated lids with slits
- ◊ two thermometers
- ◊ ruler or meter stick

What is Happening?

The darker a surface is, the more light it absorbs and the faster it heats up. The relative ability of different surfaces to absorb light is called absorptivity: the higher a surface's absorptivity the more light it absorbs. A completely black surface absorbs 100 percent of the light that strikes it. A completely white surface reflects 100 percent of the light that strikes it.

As surfaces absorb light, they heat up and give off increasing amounts of heat. In doing so, these surfaces heat the air above them. As the air heats, it becomes less dense and rises. Cooler air then moves underneath and is subsequently heated. This action creates wind and is partially responsible for sea breezes. When the sun rises, the land heats faster than the ocean. Because the land is warmer than the surface of the ocean, the air above the land becomes warmer than the air above the ocean. The warmer air rises, and air from above the ocean moves in to replace it, creating the ocean (onshore) breeze. At night, the land cools faster than the surface of the ocean, leaving air above the land cooler than air above the ocean. Air then will be rising faster above the ocean, and cooler air from above the land will move in to replace it, creating a land (offshore) breeze. This process of warm air rising and cooler air replacing it is known as convection.

Important Points for Students to Understand

- ◊ Dark surfaces absorb more light and heat more rapidly than light surfaces.
- ◊ Ground surfaces can heat the air above them.
- ◊ Differential heating of the ground leads to wind.

Time Management

This activity is easily done in one class period.

Preparation

The cups, lids, and thermometers for this activity can be bought from scientific supply companies as kits or as separate pieces. The cups are generally sold as black and silver. Silver cups work as well as white ones. A cheaper alternative is to make the materials yourself by spray painting metal cups or cans—one black and one white. Most craft stores sell 1.25 centimeter-thick styrofoam sheets. Using one of the cups as a kind of cookie cutter, cut the insulated lids out of the styrofoam. Then cut slits in the lids to insert the thermometer. Make sure the slits are large enough so that students won't have to force the thermometers through the lids. Use alcohol-filled thermometers for this activity, and *urge students to use caution to avoid breaking the fragile thermometers.*

If you are using a light source other than a 100 watt bulb, it is important that you try this activity first to determine which distances from the light give the best results.

Suggestions for Further Study

If the cups are left long enough in the light, they will eventually reach the same temperature, but it may take a very long time. Have the students explore why this is not likely to happen on Earth with adjacent light and dark surfaces.

Have students investigate the term "snow blindness" and how it relates to this activity. Have students investigate the panels used in solar heating. These panels use very dark surfaces to absorb the maximum amount of sunlight. Students can explain how people in different areas use color to adapt to their climate.

The high absorptivity of dark surfaces can explain why densely populated cities often have higher temperatures than the suburban and rural areas surrounding them. This is partly due to the heavy concentration of asphalt roads (also known as "blacktop") in cities as compared to suburban and rural areas. Encourage students to investigate this phenomenon by locating such cities and explaining the temperature differences.

Answers to Questions for Students

1. The temperature of the black cup rose faster. Answers will vary with students.
2. A dark surface heats most quickly because dark surfaces have higher absorptivity.
3. A white shirt, because it would reflect the sunlight rather than absorbing it. By reflecting sunlight, the white shirt keeps the person cooler.
4. The paved areas and plowed fields are dark surfaces. Since the air above them is likely to be rising on a sunny day, it could help the glider stay aloft longer.
5. As sunlight strikes the two surfaces, the dark one will heat faster. The air above the dark surface will heat and then rise. As the warm air rises, cooler surrounding air will move in to take its place creating a breeze.
6. A black or dark-colored roof for cold climates; a white or light-colored roof for hot climates.

Wind Elegy

(W.E.W.)

Only the wind knows he is gone,
Only the wind grieves,
The sun shines, the fields are sown,
Sparrows mate in the eaves;

But I heard the wind in the pines he planted
And the hemlock overhead,
"His acres wake, for the year turns,
But he is asleep," it said.

Sara Teasdale

Up, Up, and Away!

Background

We all have experienced clear, cold nights followed by sunny, warm days. As the daylight comes, the temperature of the Earth's surface and the air above it begin to rise. The coldest temperatures typically occur just before dawn; the highest temperatures for the day occur in mid-afternoon. What happens to air when it is heated during the day or cooled during the night? How does this heating and cooling of the air relate to weather?

In studying the effects of changing temperatures on air, the concept of density is important. Everything is made up of atoms and molecules, and the density of something just tells us how tightly packed together the atoms and molecules are. Styrofoam, for example, is not very dense; its molecules are not very tightly packed. In comparison, lead blocks are much more dense; its atoms are packed very closely together. Because they have different densities, a piece of Styrofoam that weighs the same as a piece of lead would have to be much larger.

Air is even less dense than Styrofoam. When the temperature of air changes, so does its density. This means that when the temperature of air changes, the molecules either move closer together or farther apart. In this activity, you will investigate how temperature affects air's density.

Objective

The objective of this activity is to investigate the effect of temperature on the density of air.

Materials

For each group of students:

- ◊ balloon
- ◊ empty 475 ml or 600 ml glass soda bottle
- ◊ bucket of ice water
- ◊ bucket of hot water
- ◊ safety goggles for everyone

Procedure

Trial 1

1. Place the uncovered bottle in the bucket of hot water for three minutes. Do not submerge the bottle or allow water to get into the bottle. You will probably have to hold it in place to keep it upright.
2. Place the balloon over the mouth of the bottle. You have now isolated a mass of air. It is important to remember throughout this trial that the amount, or mass, of air will remain constant. In the box marked "Trial 1 Prediction" suggest what will happen to the balloon when the bottle is placed in a bucket of ice water. Explain your prediction.
3. Before you continue, note that there is a small chance the bottle may break when placed in the ice water. *Everyone should be wearing safety goggles.* Place the bottle in the

bucket of ice water. In the box marked "Trial 1 Explanation," describe what happens. Remembering that the mass of the air has remained constant, explain what has changed.

Trial 2

1. Take the balloon off the bottle and place the bottle back in the bucket of ice water for three minutes. Do not submerge the bottle or allow water to enter it.
2. Place the balloon over the mouth of the bottle. As in Trial 1, you have isolated a mass of air. Again, it is important to remember throughout this trial that the amount, or mass, of air will remain constant. In the box marked "Trial 2 Prediction," suggest what will happen to the balloon when the bottle is placed in a bucket of hot water. Explain your prediction.

TRIAL 1 PREDICTION

TRIAL 1 EXPLANATION

TRIAL 2 PREDICTION

TRIAL 2 EXPLANATION

3. As with Trial 1, there is a small chance that the bottle may break when placed in the hot water. *Everyone should be wearing safety goggles.* In the box marked "Trial 2 Explanation," describe what happens when the bottle is placed in the hot water. Remembering that the mass of the air has remained constant, explain what has changed.

Questions/Conclusions

1. What are the two characteristics (or variables) of air that are being changed in this activity? State the relationship between these two characteristics.
2. As air is heated, what happens to its density? In other words, did the molecules of air move closer together or farther apart?
3. As air is cooled, what happens to its density? In other words, did the molecules of air move closer together or farther apart?
4. In the atmosphere, what would you expect to happen to air that is warmed? cooled?
5. Based on your observations and your answers to these questions, do you think it would be best to place a warm-air vent near the floor or the ceiling of a room? Where would you place an air conditioning vent? Explain your answers.

Up, Up, and Away!

Materials

For each group of students:

- ◊ balloon
- ◊ empty 475 ml or 600 ml glass soda bottle
- ◊ bucket of ice water
- ◊ bucket of hot water
- ◊ safety goggles for everyone

What is Happening?

The relationship between density, temperature, and volume can be somewhat confusing in meteorology. This activity is designed to help students investigate the relationship between these three important variables. A key point in this activity that students must understand is that once the balloon is placed on the bottle, the amount, or mass, of air has been isolated. (The mass of the air used in each trial remains constant.) Second, by using a bottle covered with a balloon instead of a sealed container with a pressure gauge, the pressure inside the bottle will, for the most part, be held constant and will be equal to the air pressure outside the bottle. The pressure will be held constant because the balloon allows the volume of the air to change. The two variables being manipulated in this activity are temperature and volume.

When temperature is increased, as it is in Trial 2, the volume of the air increases. Increasing volume, while mass is held constant, results in decreasing density. When air is heated it expands, becoming less dense than surrounding air and rising relative to the surrounding air. Conversely, when air is cooled it contracts, becoming more dense than surrounding air and sinking relative to the surrounding air (see Reading 5).

Many students will be familiar with the relationship between increasing temperature and increasing pressure. However, this relationship assumes that volume is held constant, which is not the case in the atmosphere. Within the atmosphere, when air temperature is increased, the expanding air mass rises creating a region of low surface air pressure. When air temperature is decreased, the contracting air mass sinks, creating a region of high pressure. This relationship between temperature and the pressure of an air mass is counter to most people's understanding of temperature and pressure. It is important to make this distinction clear. What is true in this activity and in the atmosphere is that as air is heated, it becomes less dense; and as air is cooled, it becomes more dense.

Important Points for Students to Understand

◊ As the temperature of a given mass of air increases, its volume increases and its density decreases. As the temperature of a given mass of air decreases, its volume decreases and its density increases.

◊ Less dense air rises relative to surrounding air. More dense air sinks relative to surrounding air.

Time Management

This activity requires less than 15 minutes to complete and works well as a station with other activities dealing with air pressure.

Preparation

Aside from securing sources of hot and cold water, no special preparations are required for this activity. Using a thick-walled bottle will minimize the risk of the glass breaking when the bottle is transferred from hot water to cold and vice versa. *However, to be safe, everyone involved should wear safety goggles during this activity.* It is not essential that the bottle be of the volume specified in the Materials List. Smaller and larger bottles will also work. What is important is the size of the opening. It must be small enough for the balloon to be placed over it. Plastic bottles should not be used, however, as they can deform under sudden temperature changes, perhaps confusing students and lessening the importance of the balloon.

Suggestions for Further Study

In this activity, the students are given only an intuitive definition of density. They are not presented with the mathematical definition of "density equals mass divided by volume." Depending on the development of the students, the teacher may choose to give them this quantitative relationship and show them its applications. Challenge students to quantify the relationship between air temperature, volume, and density. Students will have to devise a procedure and assemble the necessary apparatus to carry out relevant experiments.

Have students research the lowest and highest air pressures that have been recorded on Earth. What were the weather conditions associated with each? Have students relate what they learned in this activity to hot air balloons.

Suggestions for Interdisciplinary Reading and Study

The concept of density is important in geology and oceanography as well. It is hypothesized that density differences are in part responsible for the movement of continents. In oceanography, density is responsible for the global circulation of the oceans. The effects of density are more easily observed where dense ocean water comes in contact with less dense fresh water in estuaries.

Answers to Questions for Students

1. Air temperature and volume are being manipulated. (The mass of the air and pressure are being held constant.) As temperature increases, volume increases. As temperature decreases, volume decreases. NOTE: Students may respond that the characteristics are temperature and density, and this is acceptable. However, they need to understand that density changes only because volume changes.
2. The density of the air decreases. (Because volume increased but mass remained constant.)
3. The density of the air increases. (Because volume decreased but mass remained constant.)
4. Warm, less dense air rises relative to surrounding air. Cold, more dense air sinks relative to surrounding air.
5. Heat vents go near the floor. This allows the heat to rise and circulate throughout the room. If the vent were near the ceiling, the warm air would have no place to rise and would remain trapped at the top of the room. It would be best to place an air conditioning vent near the ceiling. This would allow the cool air to sink and circulate throughout the room rather than remaining at floor level.

Who Has Seen The Wind?

Who has seen the wind?
 Neither I nor you.
But when the leaves hang trembling.
 The wind is passing through.

Who has seen the wind?
 Neither you nor I.
But when the trees bow down their heads,
 The wind is passing by.

Christina G. Rossetti

Why Winds Whirl Worldwide

Background

Do trees make the wind by waving back and forth? Of course not. The trees move *because* of the wind. Wind is created when air pressure varies from place to place. In this activity, you will have an opportunity to explore where wind comes from.

Objective

The objective of this activity is to investigate how pressure differences create wind.

Materials

For the whole class:

◊ balloon (long balloons work better than round ones)

◊ string or fishing line (5 meters)

◊ drinking straw (full length)

◊ clear tape

◊ two chairs (optional)

Procedure

1. Thread the string through the straw.
2. Tie each end of the string to chairs (see Figure below) or hold one end yourself and have someone else hold the other end.
3. Blow up the balloon but do not tie it.
4. With the help of another student, tape the balloon to the straw (see insert in the Figure below).
5. Pull the string tight and move the straw to one end of the string with the untied end facing as shown in the figure.
6. Let go of the balloon and observe what happens. Record your observations in the box marked "Trial 1 Observation."
7. Repeat the process two more times, recording your observations each time.

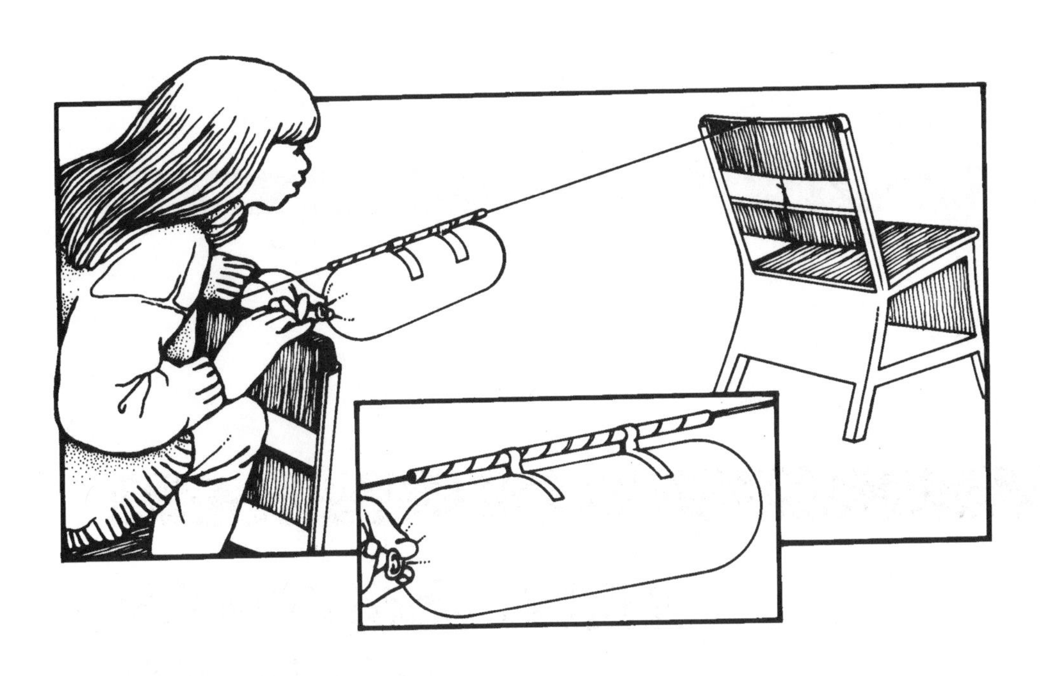

Questions/Conclusions

1. Describe what happened to the air in the balloon when it was released. Why did this happen?
2. How is what you did in this activity similar to what happens when wind is created? How is it different?

TRIAL 1 OBSERVATIONS

TRIAL 2 OBSERVATIONS

TRIAL 3 OBSERVATIONS

Why Winds Whirl Worldwide

What is Happening?

Wind results from pressure gradients, differences in air pressure from one place to another. When high pressure and low pressure areas come close to each other, air from the high pressure area will move into the low pressure area, creating wind. Because of the rotation of Earth, the air will not move directly toward the low pressure area. Instead, it spirals in, creating a cyclone. A cyclone is any weather system with winds around a low pressure area. The wind will continue until the pressure between the areas is equalized.

When a balloon is blown up, the air in the balloon is pressurized. The air around the balloon has a lower pressure than the air in the balloon. When the balloon is opened, the high pressure air rushes out to a region of low pressure. This is the same principle that governs wind.

There are, however, some differences between the model in this activity and reality. There are no solid boundaries in the atmosphere such as the boundaries of the balloon. Further, the creation of wind is usually not so dramatic in the atmosphere as with the balloon being opened. The acceleration of wind occurs gradually, due to the uneven heating that creates differences in pressure (warm air rises, cool air sinks).

Important Points for Students to Understand

◊ Wind results from pressure gradients, i.e. pressure differences from place to place.

◊ Air generally moves from an area of high pressure to an area of low pressure. (Over large distances, Earth's rotation deflects these winds to their right in the northern hemisphere and to their left in the southern hemisphere. This is known as the Coriolis effect.)

◊ Wind continues until air pressure is equalized.

Time Management

This activity will take half a class period or less.

Materials

For the whole class:

◊ balloon (long balloons work better than round ones)

◊ string or fishing line (5 meters)

◊ drinking straw (full length)

◊ clear tape

◊ two chairs (optional)

Preparation

Students often have misconceptions about wind and where it comes from. Before the lesson begins, discuss wind with your class. The object of the discussion is to reveal students' ideas of wind and its origins. Be sure that all materials are either centrally located or already distributed to the student groups.

It has been suggested that this activity may cause confusion among students because the same apparatus is used in studies of Newton's third law of motion as it applies to rocketry. The principle is identical in each situation. Gas moving from a region of high pressure to low pressure propels a rocket, and air (a gas) moving from high pressure to low pressure creates wind. Students, however, may not be ready developmentally to transfer learning across the two situations. If so, they may be confused by this activity. As an alternative, the string and straw can be removed from the activity, thus eliminating the resemblance to rocketry activities. By holding the blown-up balloon in one hand as the air is released, the students can observe that air is moving out of the balloon by holding the other hand next to the opening. The force of the air on their hand is evidence that the air is escaping. In addition, the air escaping the balloon feels like wind.

Suggestions for Further Study

Students may be interested in investigating storms, like tornadoes, in which pressure differences between two areas are very great. This would create an opportunity to talk about safety precautions that should be followed in tornadoes and strong wind storms. This activity also provides a good opportunity to talk about the land and sea breezes that were mentioned in the Teacher's Guide to Activity 7, "Which Gets Hotter?"

A discussion of the Coriolis Effect may be appropriate. Also important to airplane and rocket navigation, this phenomenon results from the rotation of Earth. Discussions of the Coriolis Effect can be found in most geology and geography textbooks.

For many years, both the Department of Energy and NASA have conducted research on wind as an alternative source of energy. The history of wind machines, modern wind machines, the economics of wind power, and the environmental aspects of wind power can be explored.

This activity naturally leads into a study of rocketry and the propulsion of rockets. While this may not be strictly related to meteorology, it is still a topic students usually find very interesting and can be used to integrate a science curriculum. The topic of waves in oceanography is also relevant to wind. Most waves in the ocean are caused by wind.

Suggestions for Interdisciplinary Reading and Study

Wind has been a source of inspiration for literature and music. Both can provide ways to introduce topics to students and provoke class discussion. Two poems about wind are included in this book. Ask students to listen for references to wind in the music they enjoy and to share these with the class.

Answers to Questions for Students

1. Most of the air left the balloon. The air in the balloon rushed out because it was at a higher pressure than the air surrounding the balloon. Some students may also correctly realize that the elasticity of the balloon is a critical factor in this particular activity.
2. Air moves from high to low pressure areas. The creation of wind is not usually as dramatic as with the balloon, and wind is not air that is escaping or is being forced from a container.

Our Hold On the Planet

We asked for rain. It didn't flash and roar.
It didn't lose its temper at our demand
And blow a gale. It didn't misunderstand
And give us more than our spokesman bargained for;
And just because we owned to a wish for rain,
Send us a flood and bid us be damned and drown.
It gently threw us a glittering shower down.
And when we had taken that into the roots of grain,
It threw us another and then another still,
Till the spongy soil again was natal wet.
We may doubt the just proportion of good to ill.
There is much in nature against us. But we forget:
Take nature altogether since time began,
Including human nature, in peace and war,
And it must be a little more in favor of man,
Say a fraction of one percent at the very least,
Or our number living wouldn't be steadily more,
Our hold on the planet wouldn't have so increased.

Robert Frost

Recycled Water: The Hydrologic Cycle

What is Happening?

Earth's supply of water is used over and over again. Water passes from ocean to air and land, and from land to ocean and air in a cyclical pattern called the hydrologic or water cycle (Figure 1). This continuous movement of water is central to meteorology.

We can trace the movement of water from the ocean to the air, from the air to the land, and back to the ocean again. In this demonstration, students will study a model of how the hydrologic cycle occurs.

Several class periods before you take up the topic of the water cycle in class, you may want to construct this model so that students can observe and informally investigate its operation.

This demonstration essentially is a distillation apparatus. As far as water is concerned, Earth is basically a closed system that can be represented by the clear shoe box. A sand bag serves as a continent, a lamp as the sun, and water as the oceans. The cup of ice represents the cold regions of the atmosphere, where water vapor condenses into water droplets.

In nature, the sun warms the water in the oceans, causing surface water to evaporate. As the water vapor rises into the atmosphere, it cools and condenses into liquid droplets. Evaporation continues, and these water droplets grow in size until they eventually fall back to the surface. Water collects on the land surface and may eventually flow back to the oceans, completing the cycle.

Objective

The objective of this demonstration is to create a model of the hydrologic cycle.

Materials for a demonstration

- ◊ clear plastic shoe box with lid
- ◊ small plastic cup
- ◊ sealed plastic bag filled with sand or soil
- ◊ water
- ◊ ice
- ◊ lamp with reflector

Important Points for Students to Understand

- ◊ The total amount of water on Earth is constant.
- ◊ Energy from the sun is the driving force behind the hydrologic cycle.
- ◊ The movement of water through the environment represents a complete cycle. There is little variation in the actual volume of ocean water.

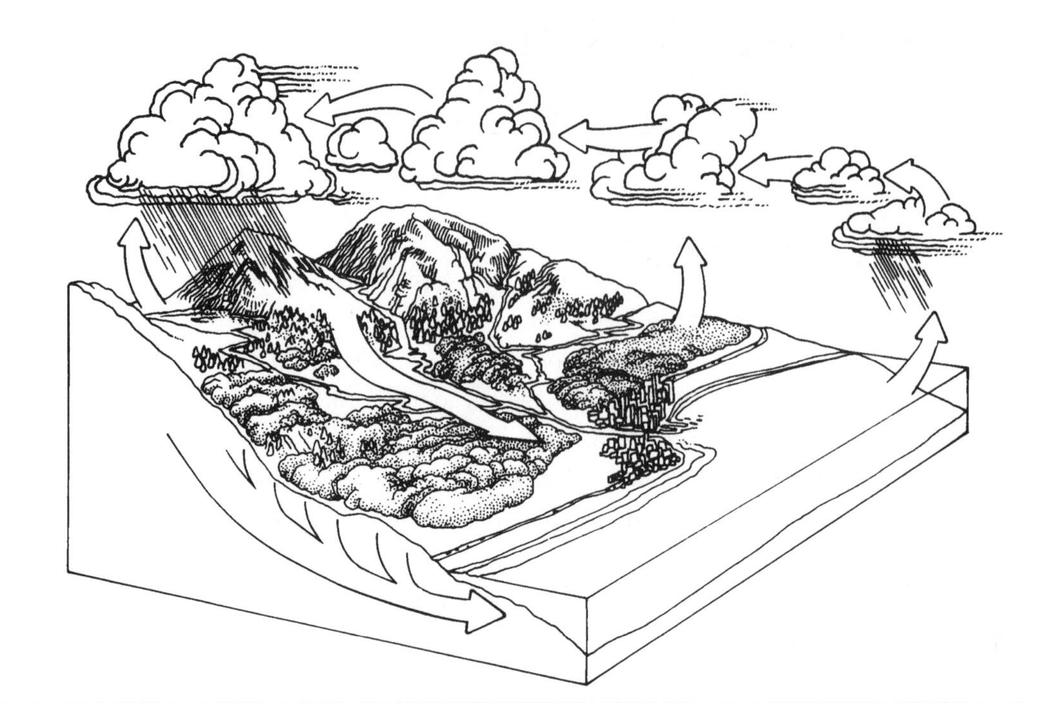

Figure 1. The hydrologic or water cycle

Time Management

Building this demonstration requires a minimal amount of time. Once the lamp is turned on, the water droplets begin forming on the cup of ice within 15–20 minutes. As long as the cup is kept full of ice, the droplets will continue to form.

Procedure

1. Set up the apparatus as shown in Figure 2.
2. Cut a hole in one corner of the lid of the clear shoe box, just large enough for the cup to fit halfway through the lid.
3. Add enough water to cover the bottom of the shoe box to a depth of about 2–3 cm.
4. Position the sand bag at one end of the box, directly under the opening for the cup. The top of the sand bag should be above the surface of the water.
5. Fill the small plastic cup with ice and place it in the opening of the lid.
6. Position a gooseneck lamp so that its light shines down onto the water inside the box.
7. Have students periodically check the demonstration to observe the progression of events. Have them record their observations over the course of the class session and diagram the movement of the water through the model.

8. Between classes, or at intervals during the day, you may want to dry off the outside of the cup and replace the ice so that each class can see the initial formation of the water droplets.

Suggestions for Further Study

Have students make diagrams representing different aspects of the hydrologic cycle.

Have students investigate what types of devices are used by countries to produce drinking water from ocean water. What issues does this bring to mind when thinking about oceanic pollution?

Suggestions for Interdisciplinary Reading and Study

This demonstration stresses the fact that the water we have on Earth is constant. There is no new water coming in from outside the planet. The concepts of the hydrologic cycle and stewardship of the water we do have are, therefore, important for students to understand. This activity demonstrates the hydrologic cycle, but students may come to appreciate it through more artistic forms. As a writing assignment, students might be asked to compose an essay on the implications for their own lives if there were no hydrologic cycle.

Activity 11, "Rainy Day Tales," is an interdisciplinary activity in itself. The activity leads students through a writing exercise in which they describe the journey of a water molecule through the hydrologic cycle.

Figure 2. Demonstration model for the hydrologic cycle

Questions for Students and Answers

1. Compare the hydrologic cycle to the demonstration apparatus. What role does the lamp serve? the sand bag? the ice?

   The lamp represents the sun and the solar energy it transmits to Earth. The sand bag represents the land. The ice provides atmospheric low temperatures necessary for cloud formation (condensation).

2. What causes water droplets to form on the outside of the cup in the shoe box?

   Condensation. The warm water vapor rises and meets the cool sides of the cup. The vapor cools and changes back to the liquid form.

3. Where in this activity does evaporation occur? condensation?

   Evaporation occurs when water in the bottom of the shoe box is heated and changes its state from the liquid phase to the gaseous phase. This vapor condenses (changes into water droplets) when it is cooled on the surface of the ice-filled cup.

4. Where in Earth's hydrologic cycle does evaporation occur? condensation?

   Evaporation: From the ocean, rivers, lakes, ponds, puddles, anywhere there is liquid water exposed to the atmosphere.

   Condensation: Primarily in clouds, but also on the ground and other surfaces as dew.

5. Was any water lost in this demonstration?

   Very little water should be lost in this demonstration, if the box is truly a closed system.

April Rain Song

Let the rain kiss you.
Let the rain beat upon your head with silver liquid drops.
Let the rain sing you a lullaby.

The rain makes still pools on the sidewalk.
The rain makes running pools in the gutter.
The rain plays a little sleep-song on our roof at night–

And I love the rain.

Langston Hughes

Rainy Day Tales

Background

Mickey the Molecule fell to Earth as part of a raindrop with many other molecules. He landed in a tree, rolled over a leaf, then dripped to the ground. Mickey had plenty of company—molecules are so tiny that he took lots and lots of molecules to make up the raindrop Mickey was in. A short time after Mickey hit the ground, the sun came out. This warmed the ground and the water molecules. Mickey evaporated back into the atmosphere and continued his travels around Earth as part of the water cycle.

You see, water molecules are pretty tough, and it isn't easy to destroy them. They are made up of three atoms: two small hydrogen atoms and one larger oxygen atom, tightly connected. But water molecules don't change their appearance when they are cooled, warmed, or come under pressure. In ice, water, and water vapor (steam) water molecules are the same.

During Mickey the Molecule's travels in the water cycle he may join other water molecules to become a water droplet in a cloud or freeze to fall to Earth as a snowflake. On Earth, he may be stored on the surface in glaciers or lakes, underground among rocks and soil, or in living things. Eventually water flows to the oceans, is transpired (given off as water vapor by plants or animals), or is evaporated directly back into the atmosphere. Throughout this cycle, the amount of water on Earth and in its atmosphere remains the same—it just changes form and does some traveling.

Objective

The objective of this activity is to explore the way water moves through Earth's environment.

Materials

For each student:

- ◊ notebook paper
- ◊ blank white paper (for illustration)

Procedure

In this activity, follow a water molecule and describe what happens to it on a trip through part of the water cycle. Select one of the following parts of the cycle:

1. river ⟶ ocean ⟶ evaporation
2. evaporation ⟶ condensation ⟶ precipitation
3. precipitation ⟶ groundwater ⟶ transpiration
4. precipitation ⟶ runoff ⟶ animal
5. snow ⟶ melting ⟶ evaporation
6. your own choice

In your story, you must do the following:

1. Have your water molecule travel through three steps of the water cycle
2. Use five action verbs from a prewriting activity
3. Write one paragraph per step; three or more sentences per paragraph
4. Illustrate your water molecule's travels

Write a quick rough draft of your story, making sure it has the components asked for. To discuss your story, your teacher will place you within a peer discussion group. After considering their comments, write a final version of your story. Be sure to polish your wording and refine your illustration.

Conclusions/Questions

1. Describe a generalized version of the water cycle.
2. How many different pathways exist through which water may cycle? Give an example of a particularly unique pathway.
3. Why is it important that we understand the water cycle? In other words, why is it important to understand that the amount of water on Earth is constant and limited?

Rainy Day Tales

What is Happening?

The water cycle describes how water moves through the environment—a fundamental concept of weather. Demonstration 10, "Recycled Water: The Hydrologic Cycle," provides a model for students to explore the changing forms of water in the water cycle. This activity combines learning this important science concept with creative writing skills, giving students a non-traditional way to consider the water cycle.

Before students begin writing, it may be useful to have a discussion about the weather patterns students are familiar with from television and radio. Discuss the origins of the moisture that falls as rain or snow, and be sure that students understand the terms: condensation, evaporation, precipitation, and transpiration. The first three terms should be fairly well established from Demonstration 10.

Materials

For each student:

- ◊ notebook paper
- ◊ blank white paper (for illustration)

Important Points for Students to Understand

- ◊ Water moves through the environment in a number of different pathways.
- ◊ The amount of water on Earth is constant and limited. Water that is used for drinking or bathing was at some point in time elsewhere in the environment.

Time Management

This activity can be completed in three class periods or less. A suggested schedule follows:

Day 1

Prewriting activity to generate a list of verbs
Begin rough draft
Homework: finish rough draft

Day 2

Peer Review: 3–4 students per group
Homework: prepare final draft and illustration
(Final draft should be in ink, on one side of the paper. Illustration should be colored.)

Day 3

Share tales on a volunteer basis.

Preparation

Ask students to read the background and procedure sections from the student materials. Discuss with students the water cycle that is depicted in the story at the beginning of the activity. Explain to students that they will be writing their own story involving the water cycle.

As a prewriting activity, direct students to list 10–20 action verbs on a sheet of paper. Some of these verbs will be used in the students' stories. Have students begin writing their story, following the guidelines listed in the procedure section of the student materials.

When students have completed their rough drafts, divide the class into groups for peer review. A peer evaluation form is included for groups to use in reviewing each others' tales. On completion of the peer review, students should prepare their final drafts and illustrations. Illustrations should be in color.

Have students share their tales on a volunteer basis and then have them turn in their final draft as well as their rough draft. A suggested scoring sheet follows.

RAINY DAY TALES—SCORING SHEET

NAME:______________________________

	POINTS POSSIBLE	POINTS EARNED
I. Three consecutive steps	10	________
II. Five action verbs	5	________
III. Illustrated cover	5	________
IV. Grammar, spelling, paragraph construction, etc.	10	________
V. Overall creativity, originality	10	________
VI. Rough draft	10	________
TOTAL POSSIBLE POINTS:	50	

TOTAL GRADE: ________

PEER EVALUATION OF RAINY DAY TALES

Author's Name: ______________________________

Evaluators' Names: __________________________

1. List all action verbs used.

2. Circle misspelled words.

3. What three steps of the water cycle does the water molecule travel through? Are they consecutive steps?

4. Is there at least one paragraph per step? Are there more than three sentences per paragraph? Are there good transitions between the paragraphs?

5. What did you particularly like about this paper? (Make at least TWO positive comments.)

6. How can the author improve her/his work? (Give at least TWO helpful suggestions.)

Suggestions for Further Study

Students may wish to write other scenarios to trace water through alternative pathways. Another way to assess their understanding of the hydrologic cycle is to present them an unlabeled version of Figure 1 and have them label the different processes in the cycle.

Have students investigate a local problem of water or air pollution and its effect on local and other water supplies. Encourage students to do research on how human activity affects the water cycle. Use newspaper weather maps to explore moisture patterns.

Suggestions for Interdisciplinary Reading and Study

This entire activity is interdisciplinary as it involves teaching science through creative writing. It is also possible to involve literature in the activity. For example, the poems "Coming Storm" by Myra Cohn Livingston (page 124) or "April Rain Song" by Langston Hughes (page 88) may be used as introductions to this activity.

Answers to Questions for Students

1. In the water cycle, water evaporates into the atmosphere and returns to Earth's surface as rain or snow. Here it may be temporarily stored in glaciers, lakes, underground reservoirs, or living things. Water then returns by the rivers to the oceans, or is transpired or evaporated directly back into the atmosphere.
2. There is an almost infinite number of pathways through which water may flow in the environment. Expect some interesting answers to the second part of this question.
3. It is important to understand the water cycle because there is a finite amount of water on Earth. This results in an interdependence between people and the environment and a need to protect the cleanliness of the water supply.

Clouds

mountains
dreaming
up the sky

Francisco X. Alarcon

A Cloud in a Jar

Background

Why do clouds form in the sky on some days but not on others? There are some specific conditions that must be present for clouds to form. First, there must be sufficient water vapor in the air. Second, the air must be cool enough to cause water vapor to condense. But these two things alone do not guarantee that a cloud will form. Something else must be present. This activity is an investigation of the third factor necessary for cloud formation.

Objective

The objective of this activity is to investigate the conditions that must be present for clouds to form.

Materials

For each group of students:

- ◊ 1 liter (or larger) clear glass jar with lid (large-mouth jars work best)
- ◊ ice cubes or crushed ice
- ◊ hot water (Caution: Even very warm water will do. Do not use water that is hot enough to burn your skin.)
- ◊ matches
- ◊ can of aerosol spray (air freshener is suggested)
- ◊ black construction paper
- ◊ safety goggles
- ◊ flashlight (optional)

Procedure

1. Fill the jar with hot water. *Do not use water that is hot enough to burn your skin.*
2. Pour out most of the hot water, but leave about 2 cm of water in the bottom of the jar. Hold the black paper upright or prop it up against some books behind the jar.
3. Turn the lid of the jar upside down and fill it with ice. Now place the lid on the jar as shown in Figure 1. Observe the jar for three minutes. If you have a flashlight, darken the room and shine the flashlight on the jar while you observe it. Record your observations in the Data Table, in the box marked "Control".
4. Pour the water out of the jar and repeat steps 1 and 2.
5. Prepare the lid so that you can immediately cover the mouth of the jar during the next step.
6. Move all loose papers away from the jar. Put on your safety goggles, then strike a match and drop the burning match into the jar. Cover the mouth of the jar immediately (with the ice-filled lid). Record your observations in the Data Table, in the box marked "Match." *Be extremely careful with the matches.*
7. Pour out the water in the jar and repeat steps 1 and 2.

Figure 1

8. Spray a very small amount of the aerosol in the jar and immediately cover the mouth of the jar with the ice-filled lid.
9. Observe what happens in the jar for three minutes and record your observations in the Data Table in the box marked "Aerosol".

DATA TABLE

Trial	Observations
Control	
Match	
Aerosol	

Questions/Conclusions

1. In all three trials of this experiment, the jar contained water vapor and cooled air. Where did each come from?
2. Did a cloud form the first time you put the lid over the mouth of the jar? How about the second and third times?
3. Look up the word "aerosol" in a dictionary and write the definition here.

   Aerosol:

4. Based on the definition given in your answer to question 3, would you classify smoke as an aerosol?
5. Based on your observations and your answers, what is the other condition—besides moisture and cool air—necessary for cloud formation?

A Cloud in a Jar

What is Happening?

Three conditions are necessary for clouds to form: sufficient moisture in the air, cooling of the air, and suspended particles in the air. The first two conditions are easy to understand, but the third requires a little more thought. Without suspended particles in the air, water vapor (moisture) will not condense to form the droplets that make up clouds. Water's surface tension keeps the tiny droplets of water vapor from sticking together. The suspended particles provide a condensation surface for the droplets. That's why they are called "condensation nuclei."

Condensation nuclei are small, usually about 0.1 micron in diameter. (A human hair is about 100 microns thick.) Common condensation nuclei include sand and dust particles, salt from sea spray, pollen, and material ejected by volcanoes. Smoke can also be condensation nuclei. That is why smog, a combination of smoke and water vapor, is so common in industrial or highly populated areas.

Important Points for Students to Understand

- ◊ Three things are necessary for cloud formation: cooling of air, water vapor, and condensation nuclei.
- ◊ Water vapor must have something to condense on in order to form the droplets that compose clouds.
- ◊ Many things can serve as condensation nuclei. Some of the most common include dust, pollen, salt from ocean spray, and smoke.

Time Management

This activity can be done in a class period or less; however, the activity should be conducted as two separate parts. The trial using matches should be completed by all students *and the matches removed* before doing the trial with the aerosol can.

Materials

For each group of students:

- ◊ 1 liter (or larger) clear glass jar with lid (large-mouth jars work best)
- ◊ ice cubes or crushed ice
- ◊ hot water (Caution: Even very warm water will do. Do not use water that is hot enough to burn your skin.)
- ◊ matches
- ◊ can of aerosol spray (air freshener is suggested)
- ◊ black construction paper
- ◊ safety goggles
- ◊ flashlight (optional)

Preparation

Before the lesson begins, discuss cloud formation with the class to determine the students' ideas on how clouds form. Ask students what they think a cloud is made of, then ask them how it forms.

Be sure that all materials are either centrally located or already distributed to the groups of students. Perhaps the students could bring clear glass jars, such as mayonnaise jars, pickle jars, canning jars, etc., from home. The jars don't have to all be the same shape, but clear glass works the best. The larger the mouth of the jar, the better the activity works.

Depending on the students, the teacher may choose to light all matches for them to reduce the risk of accidents and the temptation for horseplay. *Be careful. Flames and aerosol cans are an explosive combination.* Holding a lighted match in front of an aerosol can makes a very effective flame thrower. *Students must never have access to both the matches and the aerosol at the same time.* If in the teacher's opinion, this represents too great a risk for his or her students, it is strongly recommended that the aerosol not be used at all. The important points of the activity can still be made using only smoke.

Suggestions for Further Study

The jar in this activity represents a small cloud chamber. Building a much bigger cloud chamber would make an excellent project. Students might also try other things as condensation nuclei; for example, chalk dust.

Answers to Questions for Students

1. The water vapor came from the evaporating hot water. The air was cooled by contact with the ice-filled lid.
2. Answers depend on the student observations. There should not have been a cloud the first time. A cloud should have been observed the second and third times.
3. Aerosol—a suspension of fine solid or liquid particles in gas
4. Yes.
5. The presence of suspended particles in the air.

Snail

Little Snail,
Dreaming as you go.
Weather and rose
Is all you know.

Weather and rose
Is all you see,
Drinking
The dewdrop's
Mystery.

Langston Hughes

Just Dew It!

Background

If you watch the local news, you've probably heard the meteorologist talk about the "dew point temperature" or simply the "dew point." The dew point tells us something about how much moisture (water vapor) is in the air. There is always some moisture in the air, but the amount varies. The maximum amount of water vapor the air can hold depends on the temperature of the air.

Most of the water in the atmosphere got there by evaporation, largely from the ocean but also from lakes, rivers, ponds, and even puddles. Temperature is a major factor in determining how much and how rapidly water will evaporate from these places. When the air is cool, it cannot hold as much water vapor as when the air is warm, and therefore, less water will evaporate. Because of this, at any location, there will probably be more water vapor in the air during warm weather than during cold weather.

Now imagine that some warm air with a lot of water vapor in it begins to cool. When the air has been cooled to the point that the water vapor in it begins to form water droplets, we say that the air is "saturated" with water vapor. The temperature at which this happens is called the "dew point" temperature.

As air becomes saturated, the less it needs to cool for the vapor to condense. And likewise, the less water vapor there is in the air at a certain temperature, the more it has to be cooled to make the water vapor condense. Because of this, the dew point is a measure of the amount of water vapor in the air.

On days when the air temperature and the dew point are very close to each other, we say that the air is "humid." If, for example, the air temperature is 31°C and the dew point is 29°C, this means that there is much more water vapor in the air than a day when the air temperature is the same, but the dew point is 7 °C. We are most likely to notice water vapor on very humid, warm days, when we feel clammy or sticky because all the water vapor surrounding us prevents our perspiration from evaporating easily.

Objective

The objective of this activity is to determine the dew point of the air.

Materials

For each group of students:

- ◊ Celsius thermometer
- ◊ shiny can with top removed (aluminum cans or soup cans work well)
- ◊ water at room temperature
- ◊ ice (crushed or cubes will work)

Procedure

1. Measure the air temperature with your thermometer. Record this value in the Data Table under the column "Air Temp. (°C)".
2. Fill the shiny can half full with water. *The cans may have sharp edges. Be very careful when handling them.* Allow the can to sit for one minute. If condensation forms on the outside of the can, replace the water with warmer water until no condensation forms.
3. Place the thermometer in the can with the water as shown in Figure 1.
4. Slowly, add small pieces of ice to the can while carefully stirring (constantly, but slowly) with the thermometer. Watch the outside of the can closely for the first sign of condensation.
5. When condensation begins, immediately record the temperature of the water in the can under the column "Dew Point (°C)" in the Data Table.
6. If the temperature in the room is fairly constant, repeat steps 2 through 5 twice more, recording your data beside Trial 2 and Trial 3. Find the average dew point by adding the three individual dew points and dividing the sum by three. Record the result in the Data Table.

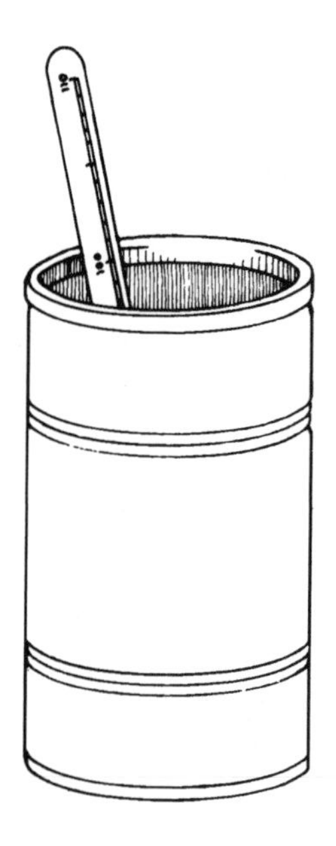

Figure 1

DATA TABLE		
Air Temp. (°C)	**Trial**	**Dew Point (°C)**
	1	
	2	
	3	
	Average Dew Point (°C)	

Questions/Conclusions

1. By how many degrees would the air have to cool in order to reach the dew point you just determined?
2. If you did this activity again tomorrow and found that the dew point had increased, would this indicate that there was more moisture (water vapor) in the air or less? Why? (Assume the air temperature is the same.)
3. When air rises, it expands and cools about 1°C for every 100 m of altitude, up to about 10 km. If the air in your room were to rise, at what height would the water vapor begin to condense and form clouds? Explain your answer. (Assume that the air in the room and the air outside are the same temperature.)
4. Under what conditions would the air temperature and the dew point be the same?
5. At what time of day are you most likely to find dew? Why?
6. Where do you normally see dew?
7. How does this activity show that dew does not fall from the sky like rain?

Just Dew It!

Materials

For each group of students:

- ◊ Celsius thermometer
- ◊ shiny can with top removed (aluminum cans or soup cans work well)
- ◊ water at room temperature
- ◊ ice (crushed or cubes will work)

What is Happening?

The capacity of air to hold water vapor depends on the air temperature. For any given temperature, there is a maximum amount of water vapor that can be present before the vapor begins to condense. The air's capacity to hold water vapor increases rapidly as temperature increases (see Figure 1).

Warm air can hold much more water vapor than can cool air. Air that is holding the maximum amount of water vapor is called saturated. The temperature at which air is saturated is identified as the dew point. When dew point is very close to the air temperature, the air's high humidity becomes easily noticeable.

In areas with many hot and humid days, the dew point is often given by meteorologists not only as a measure of the humidity but also as a kind of "comfort index." On hot days when the air temperature and the dew point are close, working outdoors can be very uncomfortable because the air is nearly saturated. When air is nearly saturated, evaporation is slowed dramatically. As a result, perspiration from our bodies is prevented from evaporating, and our cooling mechanism is impaired.

Figure 1. Air's capacity to hold water vapor varies with temperature

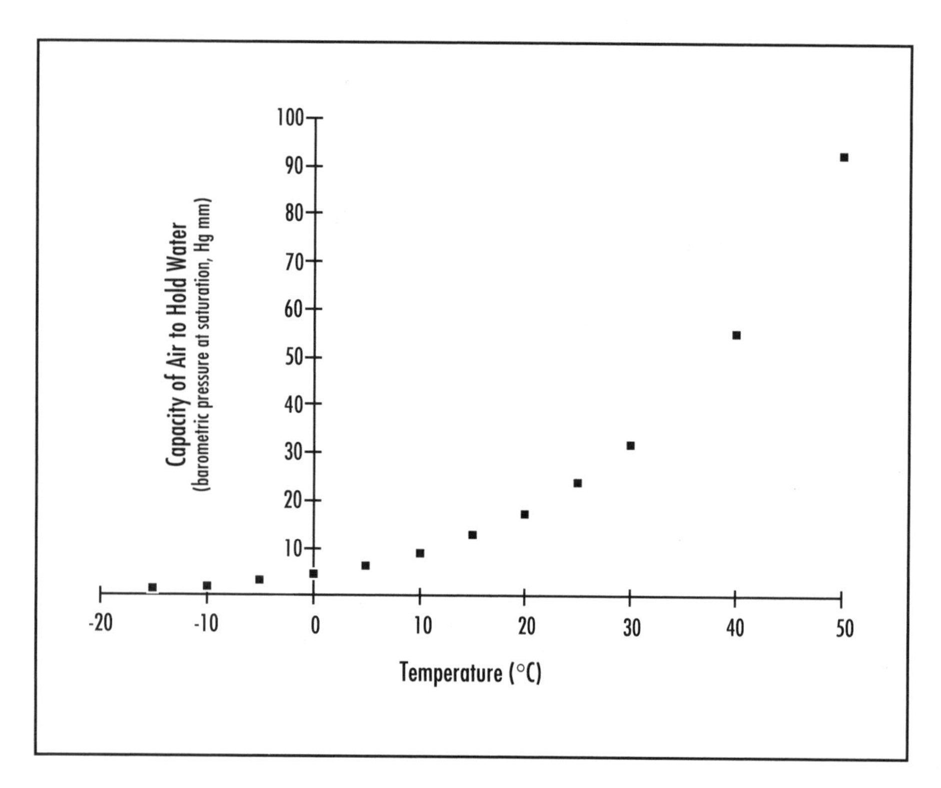

If the air temperature and the dew point are known, "relative humidity" can be calculated. Relative humidity is the ratio of the amount of water vapor *actually* in the air to the maximum amount the air could *possibly* hold at that temperature. When the dew point and the air temperature are close to each other, the relative humidity will be high. When they are the same, the relative humidity is 100 percent.

Dew forms by essentially the same process as clouds and raindrops. When moist air is cooled enough and condensation nuclei are present, water droplets will form in the atmosphere, creating clouds. Dew is the condensation of water vapor, not around nuclei in the air, but on the surfaces of objects around us.

Important Points for Students to Understand

- ◊ The maximum amount of water vapor that air can hold depends on the air temperature.
- ◊ Dew point is a measure of the amount of water vapor in the air.
- ◊ Relative humidity is a measure of the actual amount of water vapor in the air relative to the amount it can hold at its temperature. When the relative humidity is 100 percent, air is saturated.

Time Management

This activity can be completed in one class period or less. It can be repeated occasionally over the course of several weeks as local humidity conditions change.

Preparation

Before the lesson begins, discuss dew and its formation with the class to determine the students' ideas about dew.

Be sure that all materials are either centrally located or already distributed to student groups. Empty soda pop or soup cans will work well for this activity. *Because opened cans often are extremely sharp, the teacher should carefully inspect each can before use.* Use alcohol-filled thermometers for this activity, and urge students to use caution to avoid breaking the fragile thermometers.

Either crushed ice or ice cubes will work for the activity. It is somewhat easier to add small amounts of crushed ice. It is best to start with room temperature water because sometimes tap water is cold enough to cause condensation by itself.

Suggestions for Further Study

This activity leads naturally into a discussion of relative humidity. For more information on this topic, see Activity 15, "It's All Relative!" In fact, there are tables that allow you to determine the relative humidity of the atmosphere from the dew point and the air temperature. Try different methods of determining relative humidity and compare the results. Students could measure dew point and temperature and create their own tables.

Suggestions for Interdisciplinary Reading and Study

The poem "Snail" by Langston Hughes might be used to introduce the topic of dew point. Students can be asked to investigate whether or not snails actually drink dew. Also in the poem, what does the phrase "dew drop's mystery" mean?

Answers to Questions for Students

1. This depends on the students' data. The answer is obtained by subtracting the average dew point temperature from the measured air temperature.
2. More moisture. Because the air would not have to be cooled as much to reach the point where it would be saturated. This means there must be more water vapor in the air to begin with.
3. This answer also depends on the students' data. For example, if the air temperature was 20°C and the average dew point was 15°C, the air would have to rise:

   $$5°C \times (100 \text{ m} \div 1°C) = 500 \text{ m}$$

   This formula gives only an approximation. It does not take into account some factors that may influence the height at which a cloud would form.
4. In a cloud or fog. It is a misconception that it rains every time the air becomes saturated with water vapor. It is also a misconception that the air at ground level is saturated every time it rains. Often the air just beneath a raining cloud is not saturated.
5. The early morning. Because the objects on which dew collects have cooled overnight while the sun is not shining on them. These cool surfaces lower the surrounding air temperature to its dew point just as the can in this activity did.

6. Any surface that is cool enough for water vapor in the air to condense on; for example, grass, cars, metal railings, spider webs.
7. Dew formed on the can without any water "falling" from the air. It just condensed there from water vapor in the surrounding air.

Wintertime

Late lies the wintry sun abed,
A frosty, fiery sleepyhead;
Blinks but an hour or two; and then,
A blood-red orange, sets again.

Before the stars have left the skies,
At morning in the dark I rise;
And shivering in my nakedness,
By the cold candle, bathe and dress.

Close by the jolly fire I sit
To warm my frozen bones a bit;
Or, with reindeer-sled, explore
The colder countries round the door.

When to go out, my nurse doth wrap
Me in my comforter and cap;
The cold wind burns my face, and blows
Its frosty pepper up my nose.

Black are my steps on silver sod;
Thick blows my frosty breath abroad;
And tree and house, and hill and lake,
Are frosted like a wedding cake.

Robert Louis Stevenson

Let's Make Frost

Background

Where does frost come from? As you know from investigating the water cycle, water exists in three phases: solid, liquid, and gas. Water changes its phase when it is cooled or warmed.

The amount of water vapor that air can hold varies with the air's temperature. Warm air can hold more water vapor, cold air can hold less water vapor. You have probably noticed that water evaporates faster when it is warmer. The ability of air to hold water vapor is an important factor in understanding where frost comes from.

Liquid water may turn into water vapor. This is what happens when puddles evaporate. When dew forms on the grass, water is changing its phase, too. This time the change is from gas (water vapor) to liquid. Liquid water can also change into ice by freezing, and ice can turn back into liquid water by melting. Ice can also turn directly into water vapor without ever becoming a liquid by a process called sublimation. This is the same as freeze-drying, which you probably have encountered in food packaging. Another way water changes phase will be investigated in this activity.

Objective

The objective of this activity is to investigate one way that water changes phase, from water vapor to ice.

Materials

For each group of students:

- ◊ aluminum or tin can with the top removed (black cans work best)
- ◊ table salt
- ◊ crushed ice
- ◊ 2 thermometers

Procedure

1. Using your thermometer, note the air temperature. Record the value in the Data Table in the box marked "Room Temperature".
2. Pack a can with a mixture of ice and salt. There should be about twice as much ice as salt.
3. Using a pencil, make a hole in the middle of the ice. Quickly insert a thermometer into the middle of the ice (see Figure 1). *Do not force the thermometer.*
4. Watch the can carefully for the first appearance of frost. When it appears, note the temperature of the ice/salt mixture in the box marked "Trial 1" in the Data Table.
5. Empty the can and allow it to return to room temperature. Repeat steps 2–4, recording your observation in the box marked "Trial 2" in the Data Table.

Figure 1

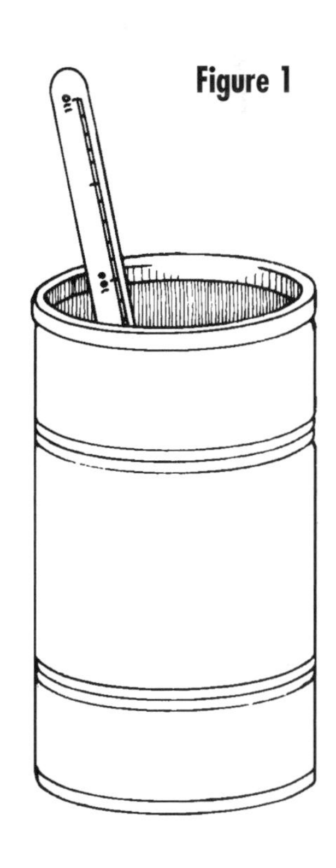

Conclusions/Questions

1. Where did the moisture come from that formed frost crystals on the can?
2. Why did the moisture deposit first as crystals of frost rather than condense as drops of water?
3. Based on your observations today and elsewhere, how are frost and dew alike? How are they different?
4. Why is it incorrect to say that frost is frozen dew? What would frozen dew look like? (Think about the phase changes involved.)
5. In a short paragraph, describe what happens when dew or frost forms.
6. With your understanding of frost formation at ground level, what might happen if the same conditions occur high up in the atmosphere?

<table>
<tr><th colspan="3">DATA TABLE</th></tr>
<tr><td colspan="2">Room Temperature (°C)</td><td></td></tr>
<tr><td rowspan="2">Ice/Salt Temperature at Frost Formation (°C)</td><td>Trial 1</td><td></td></tr>
<tr><td>Trial 2</td><td></td></tr>
</table>

Let's Make Frost

What is Happening?

On cold winter mornings, students may have seen frost on the ground or on their windows. This activity is designed to explain where frost comes from.

In the water cycle, water vapor can condense to form liquid water (see Activity 12). Water vapor can also undergo a process called deposition, whereby the gas becomes a solid without passing through the liquid phase. This is how frost forms. Frost forms when an object becomes cooler than the air adjacent to it. With sufficient cooling, the air in immediate contact with the object becomes saturated. If the air temperature remains above the freezing point, water vapor will condense on the object as dew; if the air temperature drops below freezing, water vapor will deposit as frost.

Similarly, snowflakes are formed when water vapor within the atmosphere deposits on bits of dust or other particulates. Snowflakes grow in size as more and more water vapor deposits on the existing ice crystals. Although it is said that no two snowflakes are identical, all snowflakes are hexagonal (six-sided), reflecting the way water molecules attach to each other.

In this activity, students are given the opportunity to investigate the formation of frost and to relate it to the formation of snowflakes within the atmosphere.

Materials

For each group of students:

◊ aluminum or tin can with the top removed (black cans work best)

◊ table salt

◊ crushed ice

◊ 2 thermometers

Important Points for Students to Understand

◊ Frost, like dew, is not a form of precipitation and hence does not fall from the sky.

◊ Frost and snowflakes form when water vapor, a gas, deposits as ice crystals, a solid. This is a phase change.

◊ For frost to form, air saturated with moisture is exposed to objects that are below the freezing point of water, allowing the water vapor to deposit as ice crystals on the object.

Time Management

This activity requires approximately half a class period. When steps 2–4 are performed only once, it works well as a station with other activities.

Preparation

Before the lesson begins, discuss the formation of frost with the class to determine your students' ideas on how frost forms. Common misconceptions are that frost falls from the sky and that frost is frozen dew. These misconceptions probably come from inappropriate analogies to experiences with other types of precipitation. Frost and dew are not precipitation.

Soda pop or soup cans work well for this activity if the tops are removed with a can opener. Use care when removing the tops, and cover or remove sharp edges. The frost will be easiest to detect if the cans are painted black, but unpainted cans work very well, too. Use alcohol-filled thermometers for this activity, and urge students to use caution to avoid breaking the fragile thermometers.

Suggestions for Further Study

Challenge students to determine the conditions at which dew would form on the can, and compare the two results. Along a related line of investigation, students may be familiar with the saying that sometimes it is too cold to snow. Have students investigate this saying and determine whether or not it is true. Why is Antarctica called a desert? Exploring the differences between snow, sleet, freezing rain, and hail is a good project to synthesize and evaluate student comprehension of the processes involved.

Answers to Questions for Students

1. The frost that deposits on the can comes from water vapor in the air.
2. The temperature of the can was below the freezing temperature of water.
3. Frost and dew both form from water vapor within the air surrounding the object on which they form. For dew to form, the object is above the freezing point and the moisture condenses as water droplets. For frost to form, the object is below the freezing point and the water vapor deposits as ice crystals.
4. For dew to freeze, the water vapor would have to first condense as water droplets and freeze. Frozen dew would look like tiny frozen water droplets.
5. Frost formation occurs when an object becomes cooler than the air adjacent to it. With sufficient cooling, the air in immediate contact with the object becomes saturated. If the air temperature drops below freezing, water vapor may deposit as frost. If the air temperature is above freezing, the water vapor condenses as dew.
6. Water vapor deposits on particulates to form ice crystals that can grow into snowflakes.

Fog

The fog comes
on little cat feet.

It sits looking
over harbor and city
on silent haunches
and then moves on.

Carl Sandburg

It's All Relative!

Objective

The objective of this activity is to make a hygrometer and use it to measure relative humidity.

Materials

For each group of students:

- ◊ oatmeal carton, 1.9 liter milk carton, or shoe box (anything to support the thermometers)
- ◊ two large rubber bands
- ◊ two indoor/outdoor thermometers
- ◊ scissors or knife
- ◊ shoelace—at least 15 cm long (preferably the hollow cotton type)
- ◊ 175–235 ml cup
- ◊ water
- ◊ index card

Background

Have you ever heard the word "muggy" used to describe the weather? How about the phrase, "hazy, hot, and humid"? These phrases are used to describe times when there is considerable moisture in the air. Of course, there is always moisture in the air, but the amount varies. If air is warm and dry, it is easier for water to evaporate and enter the air as water vapor. If air is cold and already holds a lot of water vapor, much less will evaporate into the air.

At every temperature there is a limit to the amount of water vapor that the air can hold. The warmer the air is, the higher this limit is. When the air is holding all the water vapor it can, it is saturated. Usually, however, there is less water vapor present in the air than it can hold.

Relative humidity is a measure of how much water vapor is *actually* in the air compared to the amount the air could possibly hold. On a muggy day, relative humidity can be high, between 80 and 100 percent. When the relative humidity is this high, perspiration does not easily evaporate. This means our bodies' cooling mechanisms are less effective. As a result, we become uncomfortable.

Saturated air has a relative humidity of 100 percent. Clouds or fog form when, and where, the air is saturated. In desert areas there is little water to enter the air and relative humidity is low. Relative humidity is measured with an instrument called a sling psychrometer or with a hygrometer.

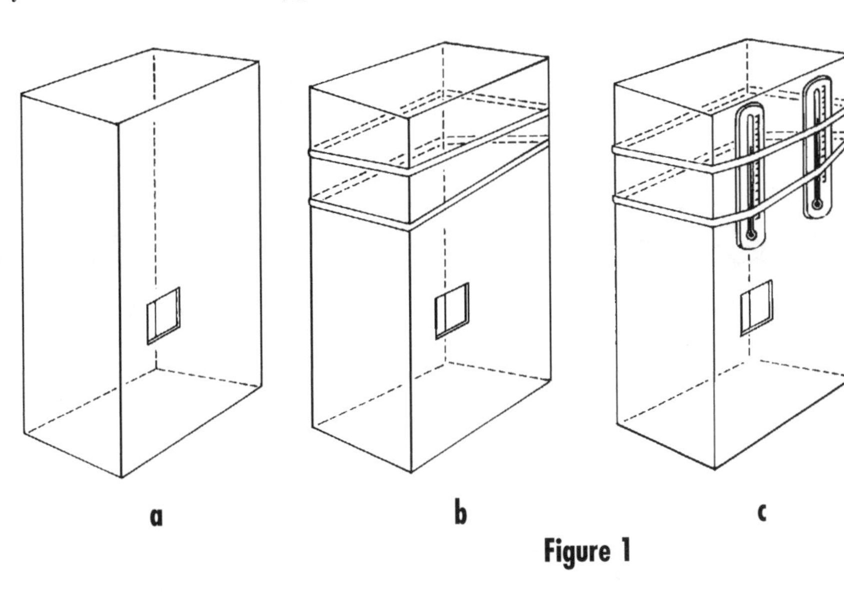

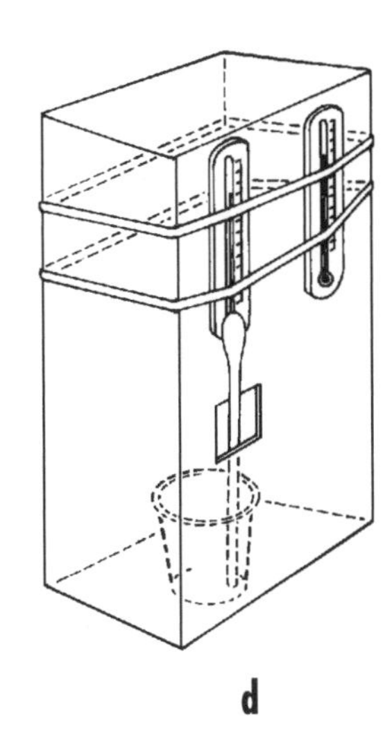

Figure 1

Procedure

1. Cut a 3 cm x 3 cm square hole in the carton or box about 12 cm from the base as shown in Figure 1a.
2. Put the rubber bands around the box as shown in Figure 1b.
3. Slide the thermometers behind the rubber bands and move one above the hole as shown in Figure 1c.
4. Fill the cup with water (almost to the top) and place it in the box as close as possible to the hole.
5. Cut a 15 cm piece of shoestring and moisten it with water. Slide one end over the bottom of one thermometer and put the other end in the cup of water as shown in Figure 1d.
6. To find the relative humidity:

   a. Read the temperature of the thermometer without the shoestring on it. Call this the "dry bulb temperature" and record it in the Data Table.

   b. Fan both thermometers with the index card for three minutes and record the lowest temperature reached by the thermometer with the shoestring on it. Call this the "wet bulb temperature" and record it in the Data Table.

   c. Subtract the wet bulb temperature from the dry bulb temperature and record the difference in the Data Table.

   d. From your calculation of the difference between dry and wet bulb temperatures, determine the relative humidity using Table 1. Record the relative humidity in the Data Table.
7. Using this method, determine the relative humidity both inside your classroom and outdoors. For each measurement, record: the location and time, both the wet and dry bulb temperatures, the difference between the two, and the relative humidity in the Data Table.

DATA TABLE

Location	Time	Dry Bulb Temp. (°C)	Wet Bulb Temp. (°C)	Difference (°C)	Relative Humidity

Table 1. Percent Relative Humidity

To determine relative humidity, find the column for the observed dry bulb temperature, then move down the column to the row for the difference between the observed dry and wet thermometers. The number in that square is the percent relative humidity. For example, if the dry bulb temperature is 20°C and the difference between the two temperatures is 10°C, the relative humidity is 24 percent.

Dry Bulb Temperature (°C)

Difference between Dry Bulb and Wet Bulb Temperature (°C)

	5	6	7	8	9	10	11	12	13	14	15	16	17	18	19	20	21	22	23	24	25	26	27	28	29	30	31	32	33	34	35
1	86	86	87	87	88	88	89	89	90	90	90	90	90	91	91	91	92	92	92	92	92	92	92	93	93	93	93	93	93	93	94
2	72	73	74	75	76	77	78	78	79	79	80	81	81	82	82	83	83	83	84	84	84	85	85	85	86	86	86	86	87	87	87
3	58	60	62	63	94	66	67	68	69	70	71	71	72	73	74	74	75	76	76	77	77	78	78	78	79	79	80	80	80	81	81
4	45	48	50	51	53	55	56	58	59	60	61	63	64	65	65	66	67	68	69	69	70	71	71	72	72	73	73	74	74	75	75
5	33	35	38	40	42	44	46	48	50	51	53	54	55	57	58	59	60	61	62	62	63	64	65	65	66	67	67	68	68	69	69
6	20	24	26	29	32	34	36	39	41	42	44	46	47	49	50	51	53	54	55	56	57	58	58	59	60	61	61	62	63	63	64
7	7	11	15	19	22	24	27	29	32	34	36	38	40	41	43	44	46	47	48	49	50	51	52	53	54	55	56	57	57	58	59
8				8	12	15	18	21	23	26	27	30	32	34	36	37	39	40	42	43	44	46	47	48	49	50	51	51	52	53	54
9						6	9	12	15	18	20	23	25	27	29	31	32	34	36	37	39	40	41	42	43	44	45	46	47	48	49
10									7	10	13	15	18	20	22	24	26	28	30	31	33	34	36	37	38	39	40	41	42	43	44
11											6	8	11	14	16	18	20	22	24	26	28	29	31	32	33	35	36	37	38	39	40
12														7	10	12	14	17	19	20	22	24	26	27	28	30	31	32	33	35	36

Questions/Conclusions

1. What did you notice about the wet bulb temperature as you fanned the thermometer with the index card? How can you explain what you observed?
2. Was there a difference between the classroom and outdoor relative humidity? If there was, how could you explain the difference?
3. If relative humidity was measured throughout the day, was there a difference in the indoor relative humidity as the day progressed? The outdoor relative humidity? How could you explain the differences?
4. Some people's hair curls when the relative humidity is high. Can you think of a way to use this fact to measure relative humidity?
5. Do water puddles evaporate faster on days when the humidity is high or on days when the humidity is low?
6. Sometimes people say phrases such as, "It's the humidity that makes us feel so hot, not the heat." What do they mean?

It's All Relative!

Materials

For each group of students:

- ◊ oatmeal carton, 1.9 liter milk carton, or shoe box (anything to support the thermometers)
- ◊ two large rubber bands
- ◊ two indoor/outdoor thermometers
- ◊ scissors or knife
- ◊ shoelace—at least 15 cm long (preferably the hollow cotton type)
- ◊ 175–235 ml cup
- ◊ water
- ◊ index card

What is Happening?

As discussed in Activity 13, "Just Dew It!," the amount of water vapor the air can hold is determined by the temperature of the air (see Figure 1). The warmer the air, the more water vapor it can hold. When the maximum amount of water vapor is present for a given temperature, the air is said to be saturated, and the water vapor will begin to condense as clouds, fog, or dew. When the amount of water vapor actually in the atmosphere is measured and divided by the amount of water vapor the air could hold at that temperature, relative humidity is obtained. The same percent relative humidity at two different temperatures actually represents different amounts of moisture present in the air. What relative humidity measures is the *actual* amount of water vapor in the air relative to the maximum *possible* amount of water vapor the air can hold.

High relative humidity has some noticeable consequences for humans. Some people's hair will curl. Static electricity is reduced. On warm, humid days sweat does not readily evaporate, thereby hindering the body's cooling mechanism. This last consequence has led to the saying: "It's not the heat; it's the humidity." Likewise, there are consequences of low humidity. When such a condition persists for a long period of time, brush and forest fires are more likely. Also, our skin becomes dry and chapped, and our clothes are more subject to "static cling." Mouth and nasal membranes dry out, and this may make it easier for viruses to enter our bodies. Some have suggested that this is why we are more likely to have colds during winter.

Important Points for Students to Understand

- ◊ Air is saturated with water when all the water vapor that air can hold at a given temperature is present.
- ◊ The amount of water vapor that the air can hold increases with the temperature of the atmosphere.
- ◊ Relative humidity is calculated with the formula:

$$\text{Percent Relative Humidity} = 100 \times \frac{\text{Amount of water vapor actually in air}}{\text{Maximum amount of water vapor air at that temperature can hold}}$$

- ◊ Humidity affects us in observable ways.

Time Management

The time consuming parts of this activity will be constructing the hygrometer and moving outdoors. Taking the measurements requires only a few minutes.

Preparation

Before the lesson begins, discuss humidity with the class. All materials should be either centrally located or already distributed to the groups of students. Cutting the hole in the box might best be done by the teacher. It is not easy to do. The simplest way to cut the holes is with a utility knife or a similar knife. *These knives are very dangerous and should be handled only by the teacher.* Use alcohol-filled thermometers for this activity, and urge students to use caution to avoid breaking the thermometers.

Some advance coordination will be required if the activity is to be conducted throughout the day or over a number of days. The teacher needs to decide if every class will construct the hygrometers or only one class will make them. If only one class makes them, the other classes may use these. This may make keeping track of data collected between classes and over a period of days easier.

Figure 1. Air's capacity to hold water vapor depends on its temperature

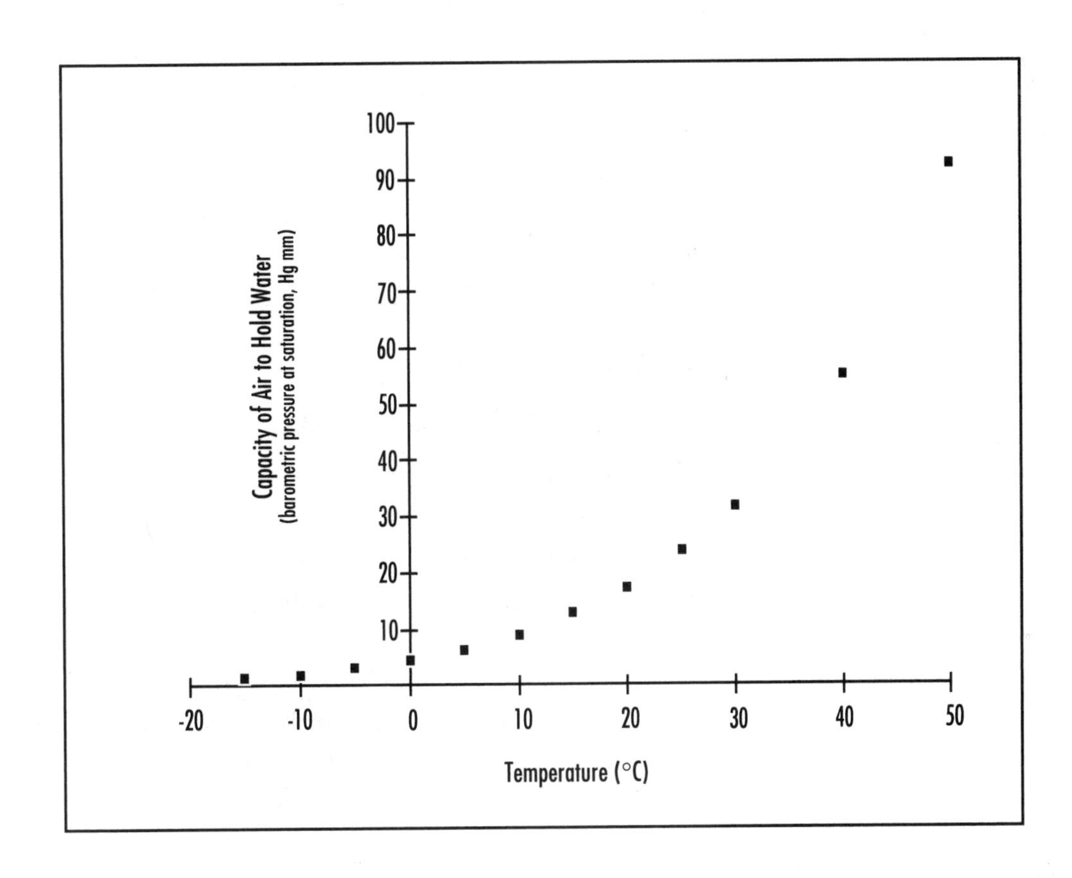

Suggestions for Further Study

The hygrometer is easily transportable. Students may find it interesting to measure relative humidity in a variety of locations to study variation. Another instructive project is to track relative humidity throughout the day and for several days, and then graph the results. Do the students see consistent trends in relative humidity as the day progresses? Do they see trends over a period of several days? It would also be interesting for them to investigate the relationship between relative humidity and the dew point. See Activity 13, "Just Dew It!"

Another instrument used to measure relative humidity is the sling psychrometer. The apparatus is similar to the one in this activity. A wet bulb and dry bulb thermometer are attached to a board. The board is attached to a string so that it can be slung in a circular motion. The slinging motion serves the same purpose as fanning the wet bulb thermometer with the index card. If students build one of these, it is essential that the thermometers be firmly attached to the board and that the string be sufficiently strong. *The sling psychrometer has a reputation for producing airborne thermometers, and this is extremely dangerous!*

Encourage students to investigate the reasons for the consequences of high and low humidity mentioned in the "What is Happening?" section of this activity.

Answers to Questions for Students

1. The temperature of the wet bulb thermometer dropped. As the thermometer was fanned, the water evaporated. As the water evaporates, it carries heat away from the thermometer and the temperature drops. (Note: Students' ability to answer this question will depend on whether or not they know that evaporation is a *cooling* process.)
2. This will depend on the students' data. In general, because of heating and air conditioning, the relative humidity will be lower indoors. Answers will vary by student.
3. Each of these answers will depend on the students' data and the weather conditions for that day. In schools with closed ventilation systems, outdoor relative humidity will be more variable, and indoor relative humidity will be more uniform. Outdoor relative humidity will generally reach a low in mid-afternoon and then begin to rise again. This trend happens because the day's highest temperatures generally occur in mid-afternoon.

4. You could just look at someone that this happens to on a regular basis and judge whether the humidity is high or not. Note: Instruments have actually been constructed to measure relative humidity based on this fact. See Figure at right.

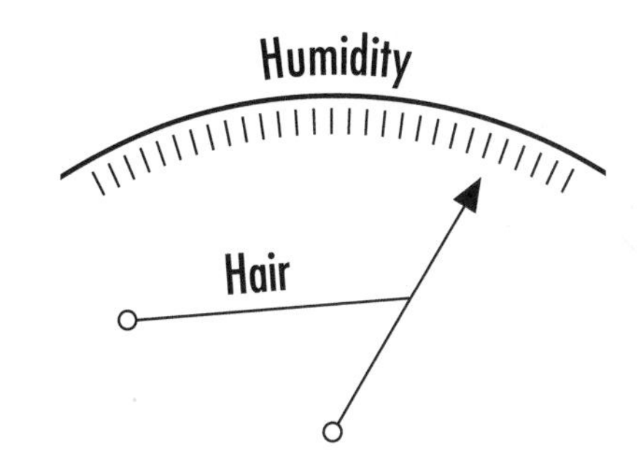

5. When other conditions are the same (e.g. wind speed, brightness of the sun, etc.), water puddles evaporate faster on days when the humidity is low. (Challenge your students to explain why and to consider what other elements affect evaporation rate.)
6. Just as high relative humidity prevents puddles from evaporating, it prevents sweat from evaporating from our skin. It is through the evaporation of sweat that our bodies are cooled. When sweat cannot evaporate, we lose the ability to cool ourselves, and we get hotter than we would in dry air. (For this reason, athletic events are sometimes cancelled when both the temperature and the relative humidity are very high. Strenuous exercise in such weather can lead to heat exhaustion, heat stroke, and other dangerous physical conditions.)

Coming Storm

You are
raising a high
roof over earth today,
shingles of gray clouds, thatches of
yellow

straw-light,
enormous folds
of wool batting unroll
as you struggle with the fierce wind
to hide

the sun,
laying dark beams
overhead, and spreading
wet tar to close up the last chinks
of blue.

Myra Cohn Livingston

Moving Masses

Background

Though all air looks the same—invisible—it has very different characteristics from place to place. Air masses are large bodies of air that have distinct temperatures and relative humidity. When air masses meet, they do not mix easily. Warmer air masses ride over cooler air masses, and cooler air masses wedge themselves underneath warmer air masses. This happens because cool air is more dense than warm air.

The boundary that forms between warm and cool air masses is called a front. As warmer air comes in contact with colder air and is forced to rise, it expands into the lower pressure found at higher altitudes. This expansion cools the air and the moisture in the air condenses, forming clouds. If the warm air continues to rise and expand, rain or snow may form.

Clouds are a visible result of the interactions of air masses. Clouds have many patterns, and by looking at cloud types it is possible to predict changes in the weather and the movement of fronts.

Objective

The objective of this activity is to investigate the types of clouds that occur when warm and cold fronts move in, and to use this knowledge to predict the weather changes over periods of a day or so.

Materials

For each student:

- ◊ scissors
- ◊ several blue crayons
- ◊ tape or glue
- ◊ paper or thin cardboard

Procedure

1. Find the page of cutouts. Separate the three strips, cutting along the dotted lines. Also cut out the "city".
2. Color the cold air blue.
3. Tape or glue the three large strips together by matching up the letters. For example, match up the letters A and tape. Then match up the letters B and tape.
4. To make the viewer itself, fold a piece of paper or thin cardboard in half lengthwise and make a vertical 6 cm cut about one-third of the way across the paper and about 2 cm above the bottom edge. Then make another cut about 7.5 cm away from the first one as shown in Figure 1.
5. Tape or glue the city below the two slits.
6. Feed the strip through the two slits. Pull from the right so that you start with Monday morning, as shown in Figure 2.

Figure 1

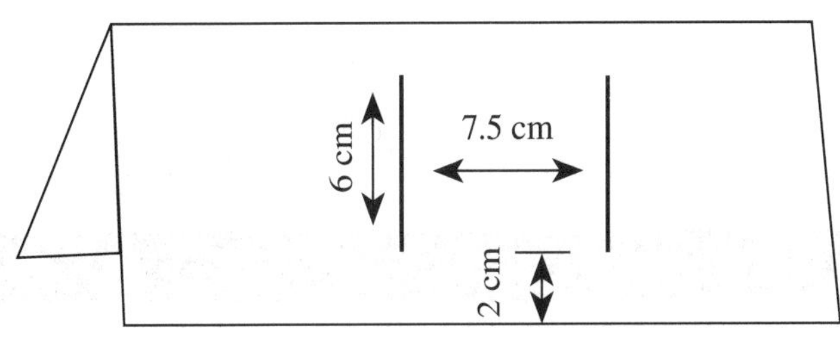

Figure 2

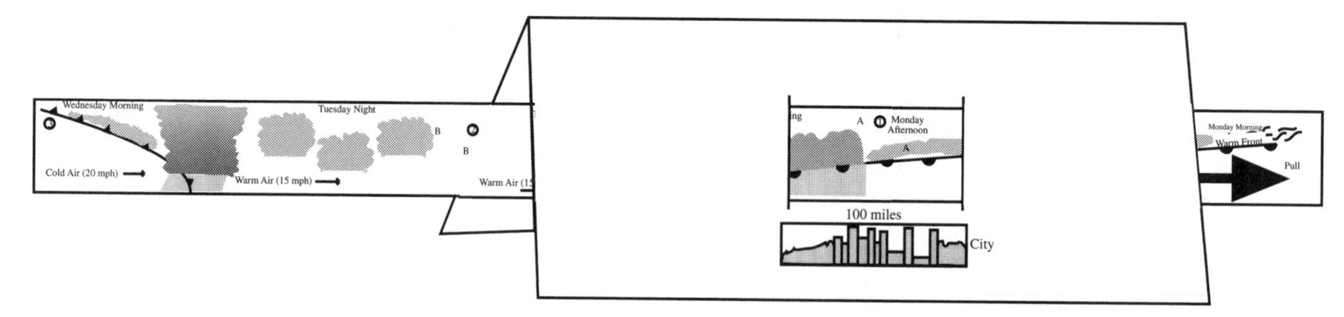

Questions/Conclusions

1. Describe the first type of cloud to appear on Monday morning.
2. What did the clouds look like by Tuesday morning when they were producing rain?
3. Why did the warm air rise up over the cold air on Monday?
4. Describe the type of clouds present as the cold front moved in on Wednesday morning.
5. If you saw thin wispy clouds followed by lower layered clouds, what type of weather might you expect in the near future?

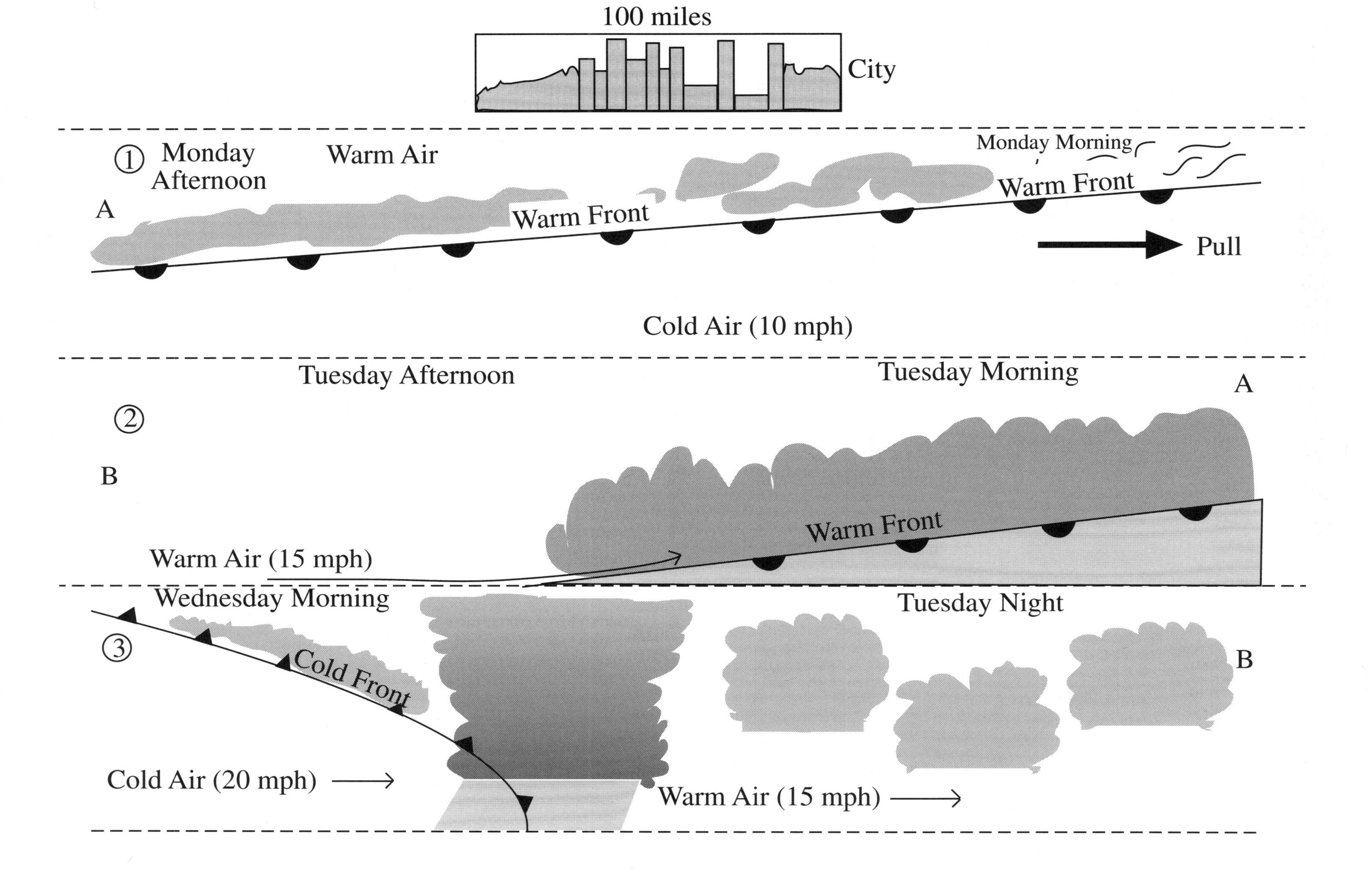

100 miles
City
① Monday Afternoon
Warm Air
Monday Morning
A
Warm Front
Warm Front
Pull
Cold Air (10 mph)
Tuesday Afternoon
Tuesday Morning
A
②
B
Warm Front
Warm Air (15 mph)
Wednesday Morning
Tuesday Night
③
Cold Front
B
Cold Air (20 mph)
Warm Air (15 mph)

Moving Masses

Materials

For each student:

- ◊ scissors
- ◊ several blue crayons
- ◊ tape or glue
- ◊ paper or thin cardboard

What is Happening?

Fronts are responsible for most of the cloudiness, rain, and snow in the United States, especially in the winter. A front is a boundary that exists between air masses of different temperature and humidity. The boundary exists because these air masses do not readily mix. The warmer air, being less dense, rides up over the cold air. As this happens, the moisture in the warm air will condense—forming clouds, rain, or snow—because the warm air expands and cools into regions of lower pressure. As warm air rises and cold air sinks, energy is released in the form of winds and a storm is born.

There are various types of fronts, and they are named according to their direction of movement. The two types discussed in this activity are warm and cold fronts. In a warm front, warm air advances replacing cooler air. In a cold front, cold air advances replacing warmer air. In each case, characteristic clouds form that allow the prediction of coming weather. The model in this activity is designed to help students recognize these characteristic types of clouds. A representation of what the cut out model might look like on a weather map is shown in Figure 1.

The cut out model is only two dimensional. Fronts and airflow are three dimensional. The fronts and their movement are actually into and out of the page in addition to across it. It is important to convey to students that the sequence of clouds presented in this activity represents an *idealized* front.

Figure 1

Important Points for Students to Understand

◊ Air masses of different temperature and humidity do not readily mix.

◊ The boundary between air masses of different temperatures is called a front.

◊ When warmer air meets colder air, the warmer air is forced to rise and cool, and the moisture in the warmer air may condense forming clouds, and eventually rain, or snow.

Time Management

This activity can be done in one class period or less.

Preparation

Be sure that all supplies are centrally located or already distributed to groups. There is a variety of posters and charts with photographs of cloud types (seethe Annotated Bibliography). These visual aids may assist the discussion of cloud types.

Suggestions for Further Study

The names for the different types of clouds were not discussed in this activity. Encourage students to investigate these on their own. Have separate classes do class predictions of the weather and compete with each other for the most accurate prediction (see Activity 1, "Weather Watch").

Students can make predictions about weather by looking at the weather maps in newspapers. Other sources of weather maps include local television news weather reports and The Weather Channel. These usually include the cold and warm fronts across the nation. This is also a good opportunity for them to learn the symbols used to represent the different types of fronts.

Suggestions for Interdisciplinary Reading and Study

Clouds can provide the motivation for a writing exercise. Most people have imagined that clouds look like people, animals, or other things. Students can be asked to write an essay describing the scene they see in the sky on a cloudy day. Myra Cohn Livingston's poem "Clouds" is an example of what such an exercise might look like in verse form.

Sayings and folklore about weather can be found in almost every culture. Some of the sayings that predict weather are based on careful observation and are fairly accurate. For example, examine the poem titled "Weather Signs," which appears at the start of Activity 1. Encourage students to read and interpret such sayings. Have them research the origin of the sayings and determine if they are based on sound principles.

Answers to Questions for Students

1. Thin, wispy, feather-like clouds (cirrus clouds).
2. Much thicker and gray.
3. It was less dense.
4. Tall, towering, billowing clouds. Various shades of white, gray, or black.
5. Warm front moving in with increased cloudiness and rain or snow likely.

Calendar Keepers

rattlesnakes
renew
themselves
each year

by shedding
their skins
by adding
a new ring

they trace
the shining
path of our
rainy seasons.

Francisco X. Alarcon

Interpreting Weather Maps

Background

Weather maps are made by combining meteorological data collected from stations all over the nation or the world. Weather stations are maintained at airports, at broadcasting stations, by schools, by private citizens, and in remote areas by the National Oceanic and Atmospheric Administration (NOAA). Weather maps usually have an outline of the area being surveyed, the names of the cities where the reporting stations are located, and symbols that represent the weather data. These weather symbols express a lot of information in a concise way. If you combine information from many stations on a map, the map will give you a picture of the large weather systems across the nation.

Figure 1 shows an example of the weather stations symbols, and the information given by each symbol. Following Figure 1 is an explanation of each type of information. As of this writing, weather station symbols in the United States are still expressed in the English system of measurement.

Objective

The objective of this activity is to learn how to interpret a basic weather map.

Materials

For each student:

◊ colored pencil

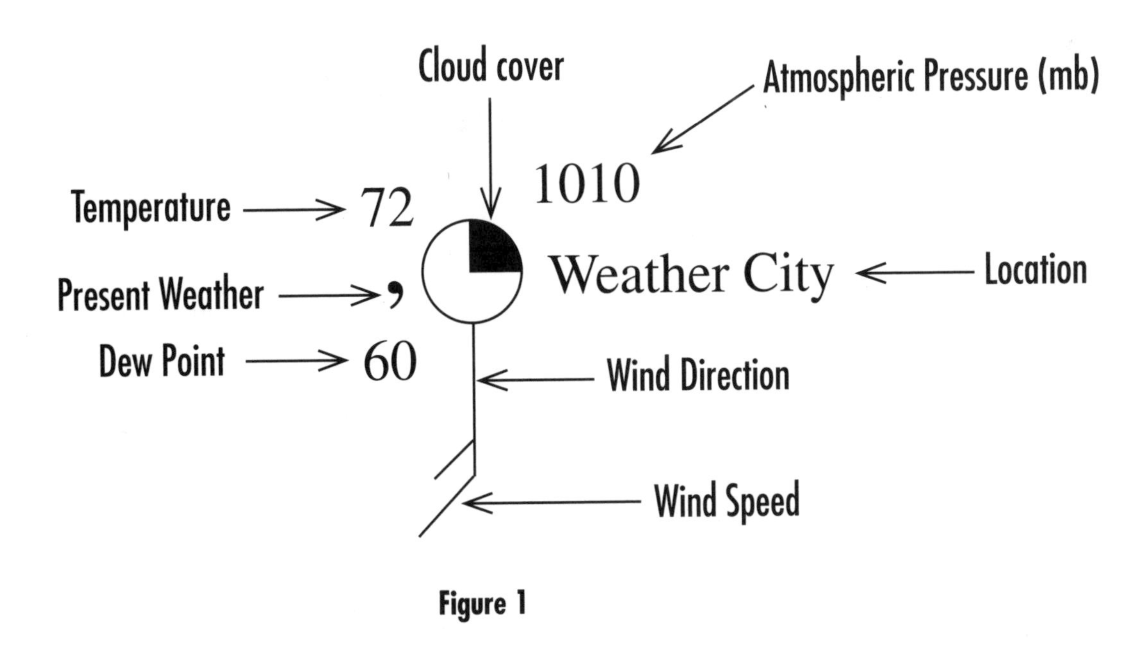

Figure 1

Atmospheric pressure: This is the atmospheric (or air) pressure measured in millibars (mb). Air pressure at sea level averages about 1013 mb (14.7 lb/in^2 or 1.04 kg/cm^2 or 760 mm Hg or 29.92 in. Hg). Often weather maps have curved lines called isobars (literally "equal bars"). These lines are drawn by connecting lines between locations on the map with the same air pressure.

Wind speed: The small lines that look like barbs represent the wind speed. Each full line represents 10 knots (kt) of wind speed (1 kt=1.15 mph=1.8 kph). Shorter lines represent wind speeds of 5 knots. Add the lines to get the total wind speed. Figure 2 shows several examples.

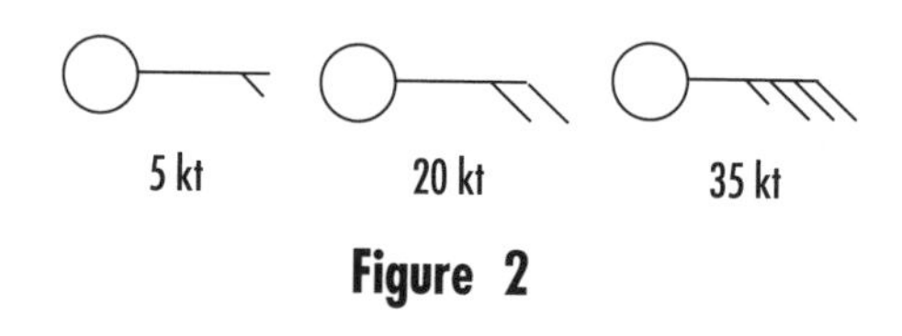

Figure 2

Wind direction: If you think of the wind speed lines as feathers on an arrow, the circle represents the arrowhead. The arrow points the direction the wind is blowing, but wind direction is designated as the direction the wind is blowing from. Therefore, if an arrow points to the east, the wind direction is actually called "from the west." In Figure 1 above, the wind direction is from the south. See Figure 3 for the principal wind directions.

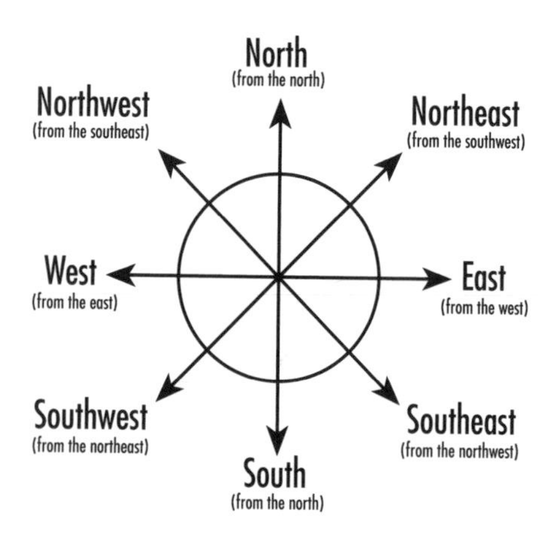

Figure 3

Temperature: This is the temperature measured in °F every hour.

Dew Point: This is the temperature in °F the air would have to be cooled to for the air to become saturated and for water vapor in the air to condense.

Cloud cover: The amount of cloud cover is represented by the amount of the circle that is blackened. Figure 4 shows some examples.

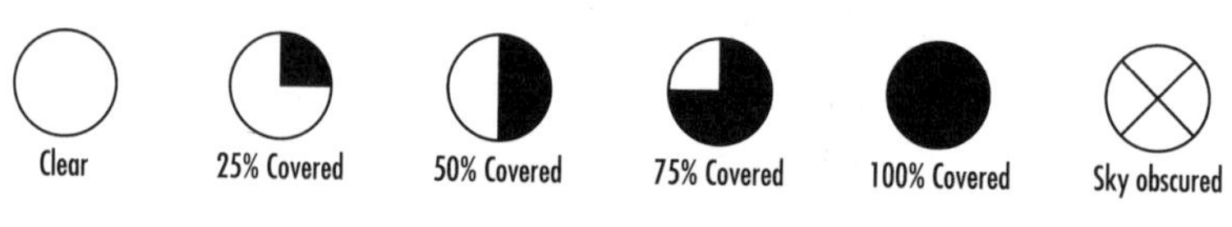

Figure 4

Present weather: Figure 5 shows a list of symbols used to designate some of the different types of weather.

Figure 5

• Intermittent rain	9 Intermittent drizzle
• • Continuous rain (light)	9 9 Continuous drizzle
△ Hail	Thunderstorms
Sleet	═ Fog
* Intermittent snow	Slight rain showers
* * Continuous snow (light)	Moderate or heavy rain showers

Procedure

1. Answer the following questions referring to the weather map in Figure 6:

   a. What is the "present" weather in Dallas, Texas?

   b. What is the atmospheric pressure in Kansas City?

   c. From which direction is the wind blowing at Hatteras, North Carolina, and what is its speed?

   d. What is the temperature in Pueblo, Colorado?

   e. What is the cloud cover in Miami, Florida?

   f. What is the atmospheric pressure in Roswell, New Mexico?

   g. What is the "present" weather in Chicago, Illinois?

   h. What is the cloud cover in New York City?

   i. From which direction is the wind blowing in Helena, Montana and what is its speed?

   j. What region of the nation appears to be generally cloudy? What region appears to be generally clear?

2. In Weather City, the atmospheric pressure is 1010 mb. The temperature is 54°F, and the dew point is 40°F. The wind speed is 15 knots from the southeast. The cloud cover is 50 percent. Draw the weather symbols that represent the data recorded at Weather City.

3. Use a colored pencil to shade lightly all areas in Figure 6 that are experiencing 100 percent cloudiness or precipitation.

4. Why is it important to be informed about weather conditions?

5. Of all the weather conditions that occur in your area, which pose threats to life and property?

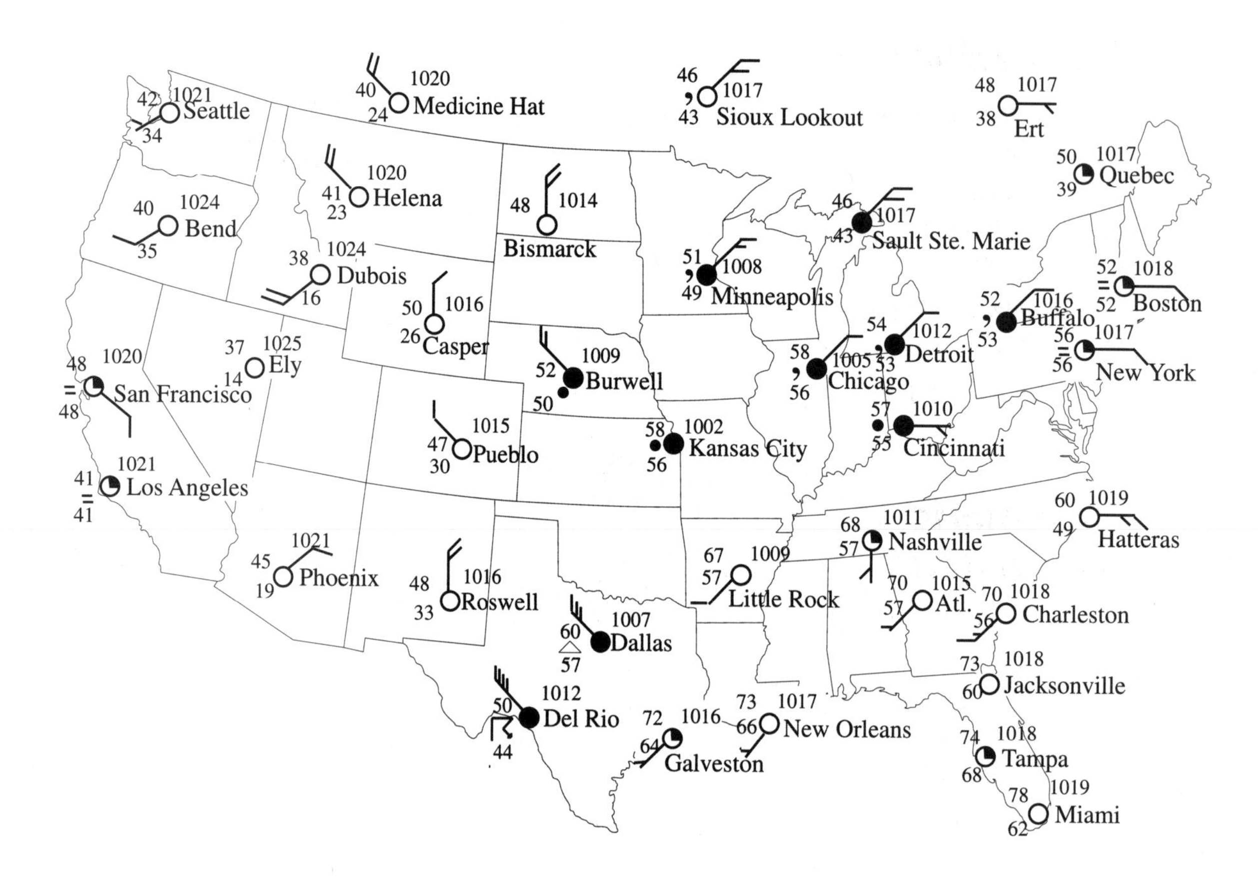

Seattle
Medicine Hat
Sioux Lookout
Ert
Quebec
Helena
Bend
Bismarck
Sault Ste. Marie
Dubois
Casper
Minneapolis
Boston
Buffalo
Detroit
New York
Chicago
Burwell
San Francisco
Ely
Pueblo
Kansas City
Cincinnati
Los Angeles
Hatteras
Nashville
Phoenix
Roswell
Little Rock
Atl.
Charleston
Dallas
Jacksonville
Del Rio
New Orleans
Galveston
Tampa
Miami

Figure 6

Interpreting Weather Maps

What is Happening?

By combining the information from many stations, an accurate picture of large weather systems can be obtained. As a result, weather maps hold an enormous amount of information about weather systems and about the movement of weather over the continent. Those skilled in weather map reading can make accurate forecasts and will be able to adjust their lives accordingly: severe weather can be anticipated, travellers can be appropriately clothed, schedules can be altered to accommodate weather changes.

Materials

For each student:

◊ colored pencil

Important Points for Students to Understand

◊ There is a network of weather reporting stations across the nation and world.

◊ The symbols used in weather maps are a form of shorthand, which makes it possible to represent a large amount of information in a small space.

◊ Each symbol has a very specific meaning.

Time Management

This activity can be completed in one class period.

Preparation

The only preparation necessary for this activity is for the teacher to become familiar with the symbols and terminology used on weather maps. The descriptions of the symbols used are provided in the student section. (The scope of this activity does not include some important but difficult to understand features of weather maps—e.g. isobars and fronts—except as suggestions for further study.)

Suggestions for Further Study

While fronts, isobars, and isotherms are difficult to understand and to construct, they are not beyond the capabilities of most middle school students. They may be interested to learn how to

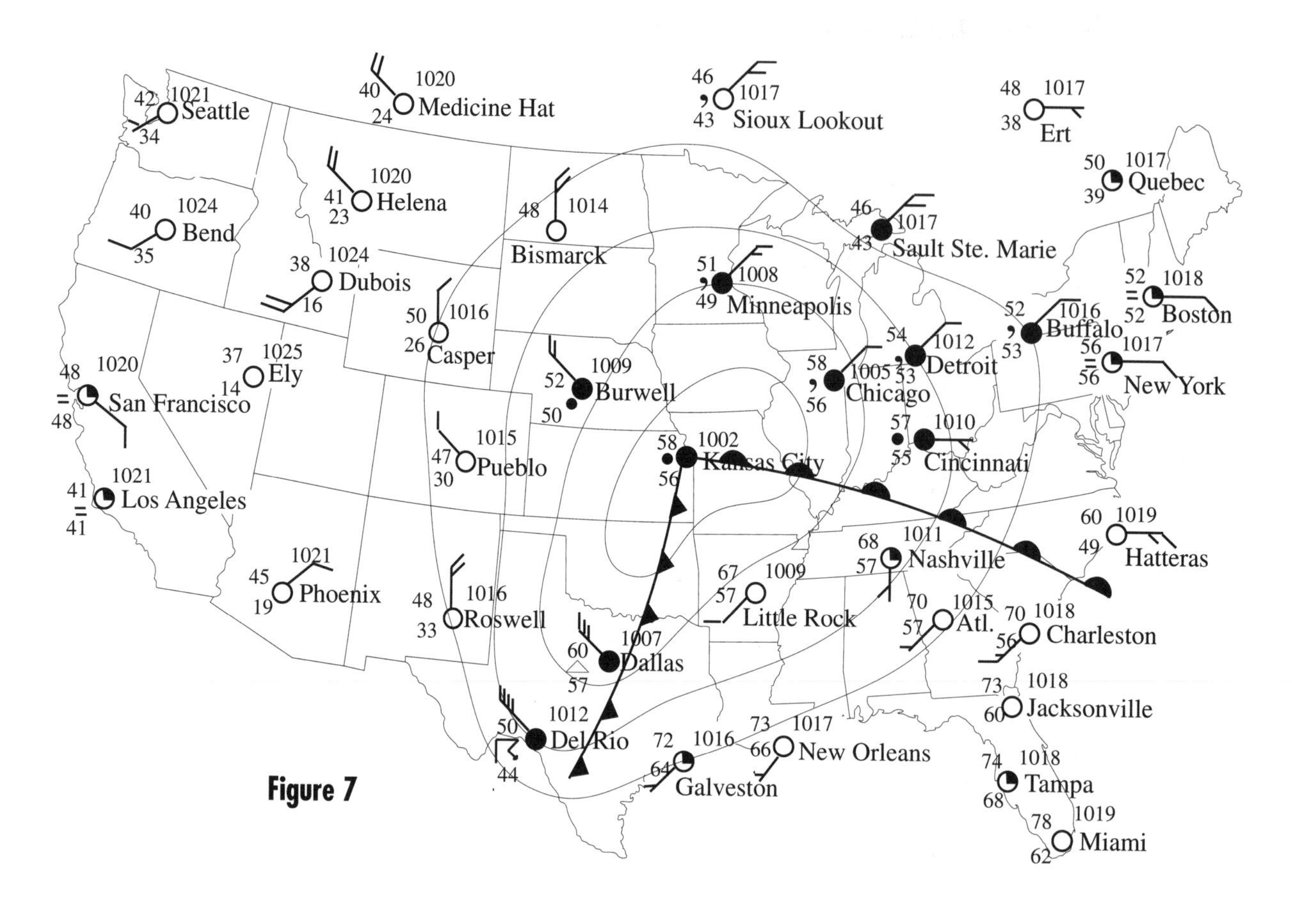

Figure 7

draw them on their own. Drawing the isobars at four-millibar intervals (996 mb, 1000 mb, 1004 mb, 1008 mb, etc.) will give them a much better understanding of pressure systems. Also, by using wind shifts and temperature changes, the fronts can be located. A diagram showing the isobars and the fronts for the weather map in this activity is given in Figure 7.

Encourage students to apply what they have learned in this activity to weather maps in the newspaper and to the ones they see on television weather reports. A topic that may interest students is weather satellites. These satellites make possible the global pictures of weather systems that are shown on television news reports.

Answers to Questions for Students

1. a. Hail
   b. 1002 mb. This is the low pressure center on the map.
   c. From the east at 15 kt.
   d. 47°F
   e. The skies are clear in Miami.
   f. 1016 mb
   g. Intermittent drizzle
   h. 25 percent covered
   i. From the northwest at 20 knots.
   j. Midwest and northeast. Most of the western U.S.
2. 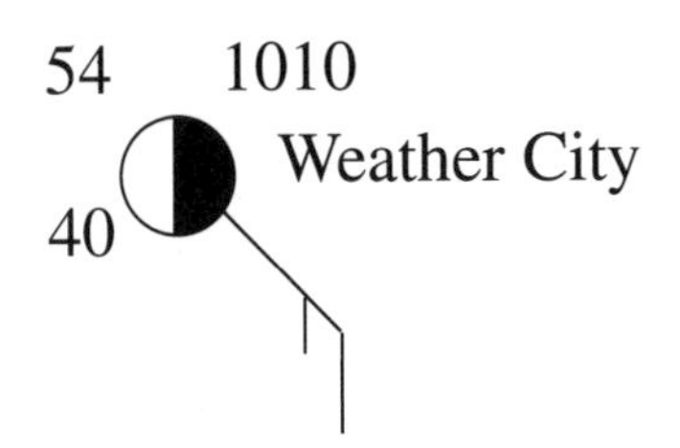

3. (Refer to Figure 7.)
4. To dress appropriately, to make travel plans, to prepare for hazardous weather, etc.
5. Answers will vary with location. For example, some parts of the nation are particularly susceptible to tornadoes. Some have dense fog on a regular basis, which may cause traffic accidents.

Hailstones

... To be reckoned with, all the same,
those brats of showers.
The way they refused permission,

rattling the classroom window
like a ruler across the knuckles,
the way they were perfect first

and then in no time dirty slush ...

Seamus Heaney

Hail In a Test Tube

Background

Every year, hail does millions of dollars worth of damage to property. A hailstorm in Denver in 1984 lasted for an hour and a half, causing over 350 million dollars of damage. Whole fields of crops have been wiped out by a single hailstorm. In the United States alone, hail causes over 750 million dollars of damage annually to crops. Hailstones can be as small as a pea, but stones larger than baseballs have also been recorded. The largest hailstone ever recorded fell in Kansas and was 14 cm in diameter, which is about the size of a softball.

Why does hail occur in some storms but not in others? Specific conditions must be present for hail to occur. In this activity you will investigate two of these conditions.

Objective

The objective of this activity is to investigate some of the essential factors in the production of hail.

Materials

For each group of students:

- ◊ crushed ice
- ◊ large test tube
- ◊ 400–600 ml beaker or similar size jar
- ◊ salt
- ◊ water
- ◊ thermometer
- ◊ stirring rod

Procedure

1. Fill the beaker three-quarters full with equal amounts of ice and cold water.
2. Pour in enough salt so that even after stirring, you can still see salt on the bottom of the beaker.
3. Wash the test tube, making sure that no dust or dirt remain inside it. Fill the test tube with cold water so that the level of water in the test tube is the same as the level of water in the beaker when the test tube is placed in the beaker. (Remember to allow for level changes in the beaker.)
4. Put the thermometer in the beaker. Then put the test tube in the beaker as shown in Figure 1.
5. Allow these to sit for ten minutes, stirring inside the beaker occasionally with the stirring rod. *Stir gently to avoid breaking the thermometer.*
6. At the end of ten minutes, remove the thermometer and record the temperature in the Data Table.
7. Remove the test tube and immediately drop a small piece of crushed ice into it. Record your observations in the Data Table.
8. Empty the test tube and repeat steps 3 through 7.

Figure 1

Questions/Conclusions

1. What is the freezing point of water in °C?
2. What was the temperature of the water in the test tube at the end of ten minutes? (Assume that the temperature of the water in the test tube is the same as the water in the beaker.)
3. What happened when the piece of ice was dropped into the test tube?
4. The water in the test tube was below water's freezing point before the piece of ice was inserted. Why do you think that the water did not freeze before the ice was inserted?
5. From this activity, what are at least two conditions that must be present for hail to form?
6. Why was it important to clean the test tube so well before you used it for this activity?

DATA TABLE

Trial	Temperature (°C)	Observations
1		
2		

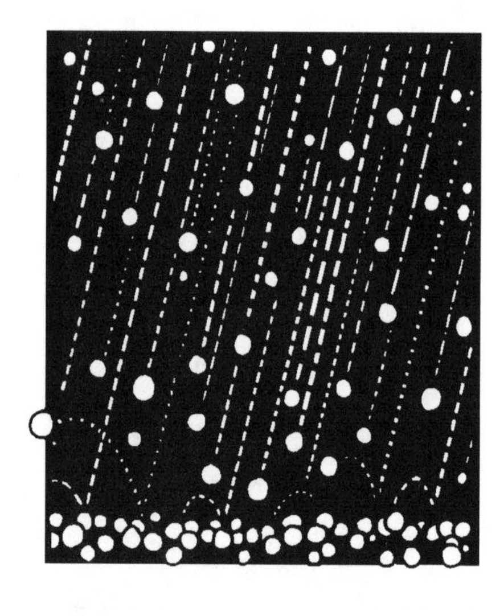

Hail In a Test Tube

What is Happening?

Hail causes millions of dollars in property damage every year. Despite its prevalence, hail requires very specific conditions to form. First, there must be supercooled water droplets in the atmosphere. Supercooled water is liquid water cooled below 0°C. In the absence of something to crystallize on, liquid water surrounded by a gas such as air can be cooled below the freezing point without freezing. Second, there must be something in the atmosphere on which the water may crystallize. Usually, this requirement is met by special particles that mimic the shape of ice crystals. Dust particles often serve this purpose. Third, there must be strong updrafts in the atmosphere. If a strong updraft exists, it can catch the small piece of ice and carry it back up into the cloud. Here more supercooled cloud droplets will crystallize on it, and it will fall again. This cycle is responsible for the growth of hailstones. The cycle will continue, and the hailstone grows in size until the force of gravity overcomes the force of the updraft.

In this activity, students will model how supercooled water can freeze instantly when a particle is added.

Materials

For each group of students:

- ◊ crushed ice
- ◊ large test tube
- ◊ 400–600 ml beaker or similar size jar
- ◊ salt
- ◊ water
- ◊ thermometer
- ◊ stirring rod

Important Points for Students to Understand

- ◊ Very specific conditions must be present for hail to form.
- ◊ In order for water to freeze, there must be a special particle present on which the ice can crystallize.
- ◊ The strength of the updraft in the cloud determines the size of a hailstone.
- ◊ Hail can be extremely destructive.

Time Management

This activity requires one class period or less to complete.

Preparation

Have all materials either centrally located or already distributed to the student groups. Be sure that ice can be acquired on the day of the activity. It is very important that the test tubes be as clean on the inside as possible. If they are not, particles may be present that will allow ice to form when the water reaches its freezing point. Use alcohol-filled thermometers for this activity, and urge students to use caution to avoid breaking the thermometers.

Suggestions for Further Study

Encourage students to try the activity with other "particles" besides the piece of ice. As an interdisciplinary project, have students research where hail occurs the most and where it does the most damage. See if students can find details about the record for the largest hailstone ever found. The record setting hailstone is mentioned in the "Background" section of this activity.

This activity might also motivate students to investigate the field of crystal growth. Encourage them to research the advantages of growing crystals in outer space, a project that NASA is currently conducting as an integral part of the space shuttle missions.

Another topic related to this activity is cloud seeding by humans to make rain. Students may mistakenly think this is a myth. The practice has gone on for decades and has become increasingly more refined. Related to this are the many different efforts at hail suppression that began in the late 1800s and were popular until the 1970s. Until the late 1970s, the United States government funded research in hail suppression. An interesting research project would be to investigate the different methods that have been used in attempts to prevent hail and why such research was stopped.

Answers to Questions for Students

1. 0°C
2. This answer will vary for each student or group of students. The temperature should be below 0°C.
3. The top portion of the water froze instantly.
4. There was nothing for the water to crystallize on. The piece of ice provided the particle on which the water could crystallize and form ice.
5. The water must be supercooled and there must be something present on which ice can crystallize.
6. To get rid of any particles capable of mimicking the shape of ice crystals that might be present in the test tube. If particles were present, the water would have frozen before the piece of ice was added.

Hurricane

Sleep at noon. Window blind
rattle and bang. Pay no mind
Door go jump like somebody coming:
let him come. Tin roof drumming:
drum away – she's drummed before.
Blinds blow loose: unlatch the door.
Look up sky through the manchineel:
black show through like a hole in your heel.
Look down shore at the old canoe:
rag-a-tag sea turn white, turn blue,
lick up dust in the lee of the reef,
wallop around like a loblolly leaf.
Let her wallop – who's afraid?
Gale from the north-east: just the Trade ...

And that's when you hear it: far and high–
sea-birds screaming down from the sky
high and far like screaming leaves;
tree-branch slams across the eaves;
rain like pebbles on the ground ...

and the sea turns white and the wind goes round.

Archibald Macliesh

Chasing Hurricane Andrew

Background

Hurricanes are the most destructive storms on Earth. They develop from tropical storms (cyclones) and are classified as hurricanes when their winds reach 64 knots (about 71 mph or 119 kph). Hurricanes include a small central region known as the eye, where the winds are light and there are few clouds. Moving out from the eye, a narrow band of intense thunderstorms, heavy rains, and strong winds is encountered. This band is called the eye wall. Beyond the eye wall are strong but diminishing spirals of the same weather. Hurricanes are huge storms. Typically they are about 500 km in diameter, and they usually last for a week or more.

Hurricanes contain tremendous amounts of energy. They gather this energy from warm ocean waters in the tropics. As the warm, humid air rises, it cools and condenses, releasing heat (called latent heat). This heat warms the surrounding air, making it lighter and causing it to rise farther. As the warm air rises, cooler air flows in to replace it, causing wind. This cooler air is warmed by the ocean, and the cycle continues. The heat from warm ocean water is the fuel that hurricanes run on. For this reason, hurricanes diminish and die when they move inland or move into colder water.

In addition to the high winds—gusts up to *172 knots* (about 192 mph or 320 kph)—and the torrential rains, hurricanes produce what is known as a storm surge. The circular winds, together with the low-pressure eye and high-pressure outer regions of a hurricane, create a mound of water in the center of a hurricane. The storm surge causes considerable flooding and is responsible for most hurricane damage and deaths.

Weather satellites in orbit above Earth can easily detect hurricanes. Satellite data, along with data from radar and aircraft, is used to follow developing hurricanes. Through tracking, we can tell where a hurricane has been. We also can estimate where it will go in the near future. When it appears that a hurricane is moving toward land, the National Weather Service (NWS) issues hurricane watches and warnings. A hurricane *watch* means that hurricane conditions are likely in the watch area within 36 hours. A hurricane *warning* means that these conditions are likely

Objective

The objective of this activity is to track the position of Hurricane Andrew for a period of six days and to distinguish between a hurricane watch and a hurricane warning issued by the National Weather Service.

Materials

For each student:

- ◊ pencil

within 24 hours. People living in low coastal areas that could be affected by a storm surge need to evacuate as soon as watches and warnings are issued.

In August 1992, Hurricane Andrew caused a tremendous amount of human suffering and billions of dollars of damage in the Bahamas, the southern tip of Florida, and parts of Louisiana. This hurricane was unusual because it struck the United States twice. After coming ashore in Florida, it passed over the Gulf of Mexico—regaining strength in the warm Gulf waters—then hit the coast of Louisiana. This activity contains the actual tracking data collected on Hurricane Andrew.

Procedure

1. Look at the data in the different parts of the table marked "The Track of Hurricane Andrew." It contains three types of information:

   a. *Date/Time*: Data was collected on Andrew every six hours beginning August 16 through August 28. Only a portion of the data is presented here. Time is given in the military convention; for example, 1200 is 12:00 noon, and 1800 is 6:00 pm.

   b. *Position*: This is the position of the eye of the hurricane by latitude and longitude. It is important to remember that the storm is much bigger than the eye. The winds extend out beyond the eye about 100 km in all directions (about one-half the area of one 5° longitude–latitude square on the map).

   c. *Wind speed*: This is the maximum speed of the winds in the hurricane, not the speed with which the hurricane is actually moving. Wind speed is given in knots (kt).
   1 kt = 1.15 mph = 1.85 kph.

2. Plot the data given in the tracking table on the map your teacher has supplied. Make a dot for each position of Andrew, and then connect the dots. For each position at the beginning of a day (time=0000), draw a small star or asterisk over the dot. You will be asked to stop plotting data periodically and issue hurricane warnings and watches based on the path of the hurricane you have plotted. REMEMBER: A hurricane *warning* means hurricane conditions are likely for a location within 24 hours. A hurricane *watch* means hurricane conditions are likely for a location within 36 hours.

Questions/Conclusions

1. Where did Andrew do the most damage before striking Florida?
2. Describe the motion of the storm displayed on your tracking map from the first point you plotted to the last.
3. What happened to the direction of Andrew after it struck Louisiana?
4. What happened to the wind speed in Andrew after it came aground in Louisiana? Why did this happen?
5. Judging from the wind speed, when did Andrew become a hurricane and when should it have been downgraded to a tropical storm?
6. In terms of the damage done, why was it so devastating for Andrew to hit the southern part of Florida? Why might it have been less destructive if it had hit farther north on the coast of the United States; for instance, Georgia or South Carolina?

Date/Time	Position Lat. (°N)	Lon. (°W)	Wind Speed (knots)
Aug 21/0000	23.2	62.4	45
0600	23.9	63.3	45
1200	24.4	64.2	50
1800	24.8	64.9	50
Aug 22/0000	25.3	65.9	55
0600	25.6	67.0	60
1200	25.8	68.3	70
1800	25.7	69.7	80
Aug 23/0000	25.6	71.1	90

STOP! Question #1: Based on how far the storm has traveled over the last 24 hours and its direction so far, for which locations would you issue hurricane warnings and watches? You can tell how far the hurricane has traveled in the last 24 hours by looking at the distance between the last two stars or asterisks you have drawn on the map. Don't forget that the size of the hurricane is much larger than the dots you have drawn.

Date/Time	Position Lat. (°N)	Lon. (°W)	Wind Speed (knots)
Aug 23/0600	25.5	72.5	105
1200	25.4	74.2	120
1800	25.4	75.8	135
Aug 24/0000	25.4	77.5	125

STOP! Question #2. Based on how far the storm has traveled over the last 24 hours and its direction so far, for which locations would you issue hurricane warnings and watches?

Date/Time	Position Lat. (°N)	Position Lon. (°W)	Wind Speed (knots)
Aug 24/0600	25.4	79.3	120
1200	25.6	81.2	110
1800	25.8	83.1	115
Aug 25/0000	26.2	85.0	115

STOP! Question #3. Based on how far the storm has traveled over the last 24 hours and its direction so far, for which locations would you issue hurricane warnings and watches?

Date/Time	Position Lat. (°N)	Position Lon. (°W)	Wind Speed (knots)
Aug 25/0600	26.6	86.7	115
1200	27.2	88.2	115
1800	27.8	89.6	120
Aug 26/0000	28.5	90.5	120

STOP! Question #4. Based on how far the storm has traveled over the last 24 hours and its direction so far, for which locations would you issue hurricane warnings and watches?

Date/Time	Position Lat. (°N)	Position Lon. (°W)	Wind Speed (knots)
Aug 26/0600	29.2	91.3	115
1200	30.1	91.7	80
1800	30.9	91.6	50
Aug 27/0000	31.5	91.1	35

STOP! Question #5. Based on how far the storm has traveled over the last 24 hours and its direction so far, for which locations would you issue hurricane warnings and watches?

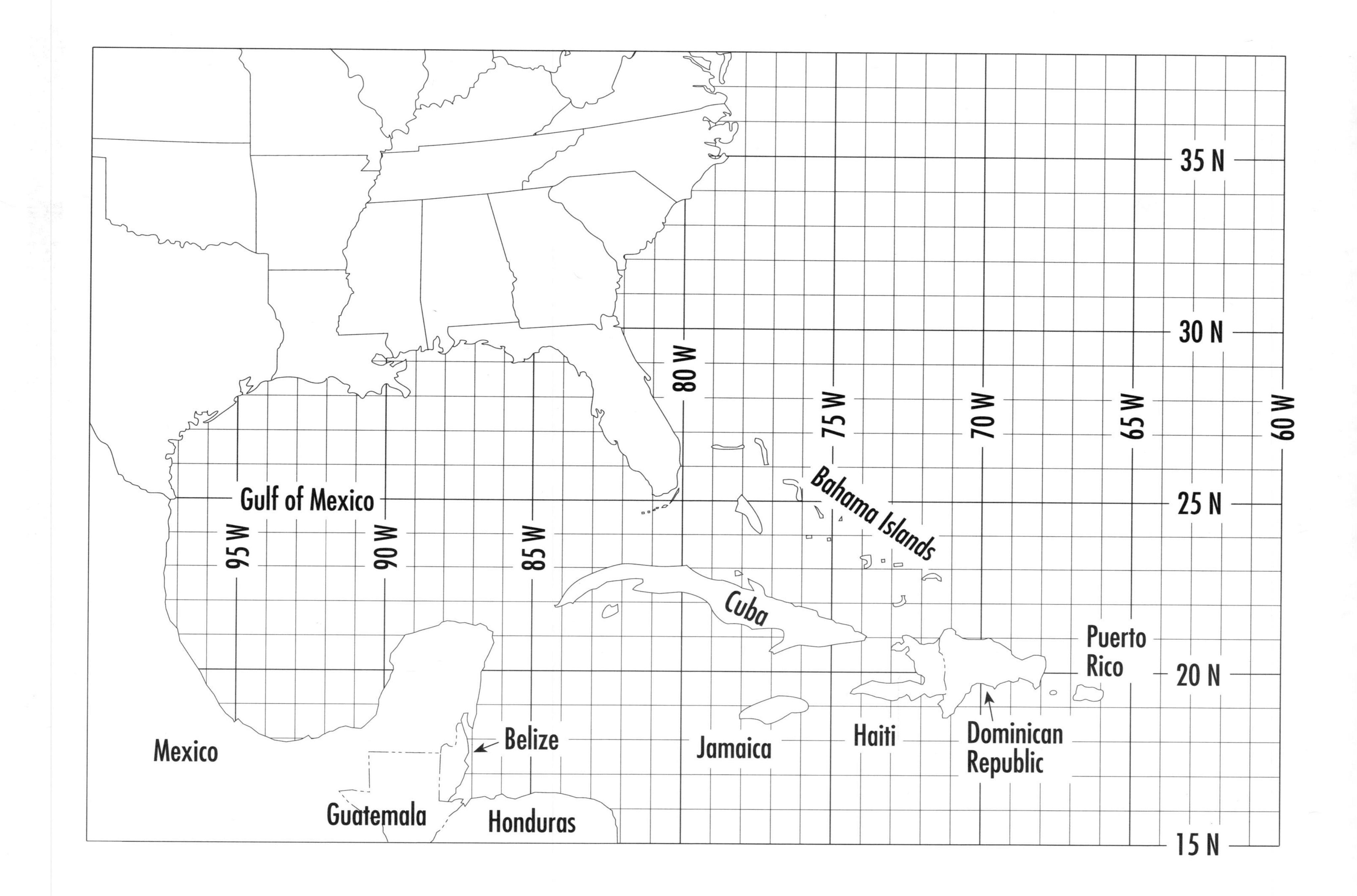
35 N
30 N
25 N
20 N
15 N
80 W
75 W
70 W
65 W
60 W
95 W
90 W
85 W
Gulf of Mexico
Bahama Islands
Cuba
Puerto Rico
Mexico
Belize
Jamaica
Haiti
Dominican Republic
Guatemala
Honduras

Chasing Hurricane Andrew

Materials

For each student:

◊ pencil

What is Happening?

This activity introduces students to the topic of hurricane tracking and the distinction between hurricane watches and hurricane warnings, which are issued by the National Weather Service. Weather satellites, along with radar and aircraft, provide a fairly accurate way of detecting hurricanes and determining their positions. The dramatic pictures of hurricanes shown on television weather reports are taken by satellites. By plotting the data gathered by these technologies, the paths of hurricanes can be determined. Combining data on a storm's previous motion with computer model forecasts of typical storm motions, meteorologists can estimate where a hurricane will go next and provide warnings to those areas. A hurricane *watch* means that hurricane conditions are likely for the specified area within 36 hours. A hurricane *warning* means that landfall is expected within 24 hours. People living in areas susceptible to storm surges need to evacuate when watches and warnings are posted. This activity will give students an idea of how hurricanes are tracked and when warnings and watches are issued.

One confusing aspect of hurricane tracking is the size of the hurricane. The coordinates given are for the eye of the hurricane only. But as mentioned earlier, hurricanes are typically 500 km in diameter. One degree on the longitude–latitude coordinate system represents about 111 km. This means that Hurricane Andrew would fill one of the 5° longitude–latitude squares on the map at any point. Hurricane force winds generally occur only within about a 100 km (1° latitude) radius of the eye. Students need to be aware of this when issuing their warnings and watches (See the Preparation section for more information on this aspect of the activity.).

Important Points for Students to Understand

◊ Satellites, aircraft, and radar can tell us where hurricanes are and give us important clues about where they are going.

◊ By tracking a hurricane and using hurricane prediction models, it is possible to predict where hurricanes will go in the near future and to issue hurricane watches and warnings.
◊ People living in areas that are likely to be covered by watches and warnings or affected by storm surges need to evacuate.

Time Management

This activity can be done in one class period.

Preparation

Before the lesson begins, discuss hurricanes and how they form. When students are predicting who should receive hurricane warnings and watches, it is essential for them to understand that hurricane winds extend far beyond the eye of the hurricane. Typically these winds extend about 100 km in all directions from the eye. This means that the area affected would be equivalent to almost half of one of the 5° longitude–latitude squares. One way to help the students visualize this is to give them an object that approximates this size on the tracking map. Depending on the map, different coins may serve this purpose, or the students can cut out circular pieces of paper. Students will have a better grasp of what locations to include by placing the object over the last place plotted prior to issuing watches and warnings.

It is helpful to have a world map or globe with longitude and latitude lines on it so that students can put Andrew in a bigger context. It is essential that students have prior experience plotting longitude and latitude data.

Suggestions for Further Study

Newspapers across the nation extensively covered the damage done by Andrew. The *Miami Herald* probably gave the most complete coverage. Encourage students to research the effects of the storm using these newspapers. Students might find it interesting to learn how hurricanes are named. They might also be interested in learning that hurricanes are referred to differently in different parts of the world or how people prepare for their arrival.

Not all tropical storms develop into hurricanes. Very specific conditions must exist. Encourage students to learn about these conditions. We usually hear most about hurricanes that hit land. What can you learn about the effects of hurricanes at sea?

Suggestions for Interdisciplinary Reading and Study

Archibald Macliesh's poem "Hurricane" provides a description in verse of a hurricane. The poem can be used to initiate a class discussion about several topics. What does the poem say about people's reaction to hurricanes? What are signs on land that indicate a hurricane is near? What signs in the ocean that can be viewed from land indicate that a hurricane is near? In the poem, how do animals announce the arrival of a hurricane in the poem?

Answers to Questions for Students

1. The Bahamas
2. For the first day, Andrew travelled northwest. For the next three days, it travelled almost due west, and then turned to the north again in the fourth and fifth days.
3. It curved almost due north and then to the northeast. Although the students would have no way of knowing this, the hurricane turned so dramatically due to an approaching cold front over the United States.
4. It slowed dramatically because the storm moved over land and lost its source of energy (warm ocean water).
5. Andrew became a hurricane between 6:00 am and 12:00 pm on August 22. It was downgraded to a tropical storm between 12:00 pm and 6:00 pm on August 26.
6. By striking the southernmost part of Florida, the storm was able to move inland, do tremendous damage, and then move back out over the warm water, where it could regain strength and make another hit. If Andrew had made landfall farther north along the coast of the United States, it probably would have moved inland and died quickly, as it did in Louisiana. This would have made for only one hit instead of two, and probably much less damage would have occurred. Another important factor is the high population density in southern Florida compared to many locations farther north.

Hurricane Warnings and Watches: Students can estimate where Andrew is going by looking at the direction and distance it has traveled in the last 24 hours and assuming that it will not deviate dramatically in the next 36 hours. They must also take into account the size of the storm. The answers provided here are based on a plot of the data in the activity, not on the actual National Weather Service warnings and watches. It would be very interesting for students to compare their answers to the watches and warnings issued by the National Weather Service. These can be obtained from a National Weather Service office. Students should have posted the following watches and warnings:

1. Hurricane warnings: Bahamas
   Hurricane watches: Bahamas, southern coast of Florida

2. Hurricane warnings: Southern coast of Florida
   Hurricane watches: None

3. Hurricane warnings: Coasts of Alabama, Mississippi, Louisiana, and Texas
   Hurricane watches: Coasts of Alabama, Mississippi, Louisiana, and Texas

4. Hurricane warnings: Coasts of Louisiana and Texas
   Hurricane watches: None

5. Andrew has been downgraded to a tropical storm.

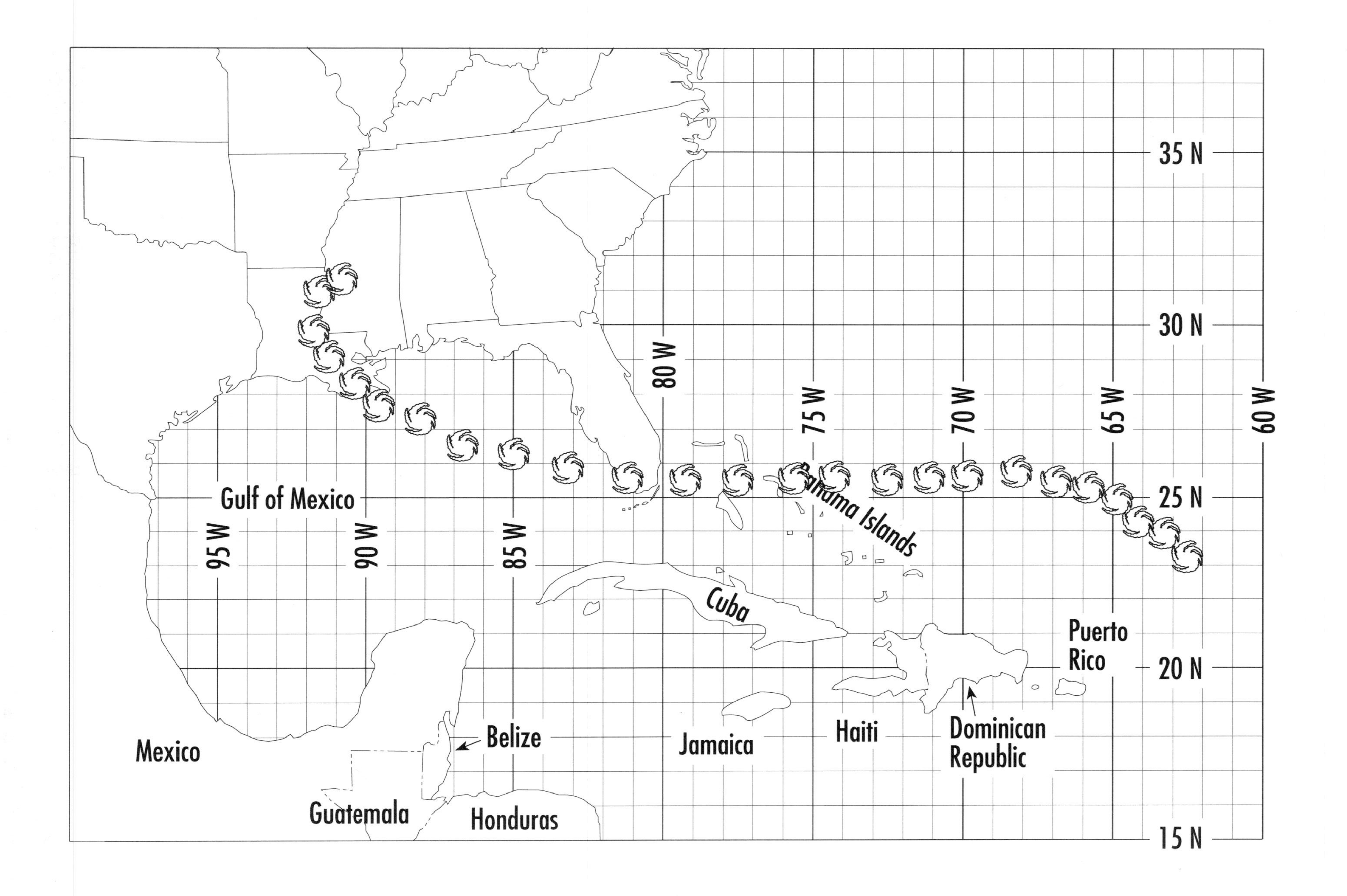

35 N
30 N
25 N
20 N
15 N
80 W
75 W
70 W
65 W
60 W
Gulf of Mexico
95 W
90 W
85 W
Cuba
Puerto Rico
Mexico
Belize
Jamaica
Haiti
Dominican Republic
Guatemala
Honduras

Readings

The readings included in this section either expand on topics presented in the activities, or provide other topics suitable for class discussion. Some are written especially for this volume; others are excerpts from scientific articles or from teacher materials. Although intended as background material for the teacher, these readings can also be used as supplemental material for students interested in extending their studies. Most of readings are connected to *sci*LINKS® for further Internet resources.

Order of Readings:

1. Earth's Atmosphere
2. El Niño Southern Oscillation (ENSO)
3. The Facts About the Ozone
4. Air Pollution and Environmental Equity
5. Weather and the Redistribution of Thermal Energy
6. Global Warming and the Greenhouse Effect
7. Environmental Effects of Acid Rain
8. Weather's Central Actor: Water
9. The Inner Workings of Severe Weather
10. Flash to Bang

Earth's Atmosphere

Earth's atmosphere plays a crucial role in shaping the planet's weather, climate, and life-supporting characteristics. It is a fairly homogeneous mixture of gases, mostly nitrogen (about 78 percent) and oxygen (about 21 percent), with suspended particles. The particles in the atmosphere come from a variety of sources including volcanoes, wind borne pollen and dust, vehicle exhaust, industrial plants, and combustion processes. The concentrations of other atmospheric gases vary widely from place to place and time to time. Variable atmospheric gases include water vapor, ozone, carbon dioxide, and pollutants such as sulfur dioxide, nitrogen oxide, and carbon monoxide.

Topic: atmosphere
Go to: www.scilinks.org
Code: PESM163

While the importance of Earth's atmosphere cannot be disputed, its size, relative to Earth itself, is minuscule. Compared with the radius of Earth (6,370 km), the depth of the atmosphere is quite shallow (120 km), amounting to a thickness of just 2 percent of the radius of Earth. Even more startling, over 99 percent of the mass of the atmosphere is restricted to a layer that is just 1 percent of the radius of Earth.

Atmospheric Levels

Temperature variations with altitude are used to identify four major layers of the atmosphere (see Figure 1). Other atmospheric characteristics (such as pressure, moisture content, and particle type and concentration) can vary widely. These variations are responsible for the ever-changing weather patterns we experience.

The lowest layer of the atmosphere, the troposphere, is characterized by temperatures that typically decrease with increasing altitude. We live in this layer, where all weather phenomena—e.g. rain, snow, hurricanes, thunderstorms, etc.—occur. Most of the mass of the atmosphere (well over 50%) is contained in the troposphere.

The stratosphere lies above the troposphere. The temperature of the lowest portion of the stratosphere is nearly constant, while the upper portion warms quickly with increasing altitude. Chemical reactions, activated by the absorption of ultraviolet radiation from the sun, lead to the establishment and maintenance of a layer of ozone (O_3) within the upper stratosphere. The ozone layer absorbs most of the sun's ultraviolet (UV) radia-

Figure 1. Earth's atmosphere has four primary layers distinguished by temperature

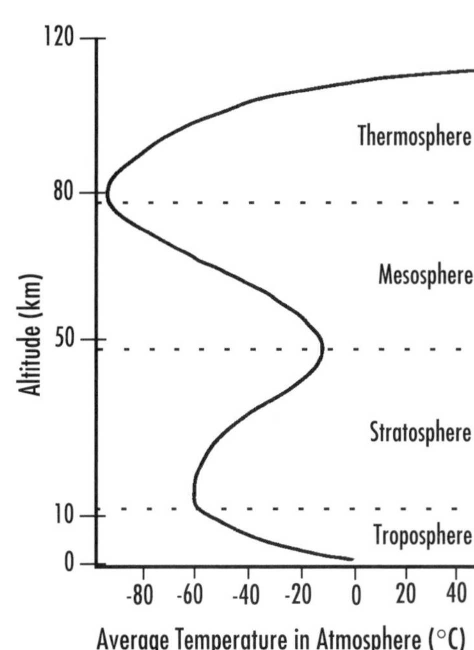

tion, resulting in increased temperatures in this region of the stratosphere. UV absorption by ozone shields Earth's surface from excessive amounts of this radiation. Overexposure to UV radiation has been shown to promote the onset of melanoma (skin cancer). Various substances introduced into the atmosphere by human activities pose a threat to the ozone layer through their chemical reactions with ozone. Chlorinated fluorocarbons (CFCs) found in some aerosol sprays and refrigerants are of particular concern because of their destructive effects on the ozone layer (see Reading 2, "Progress and Challenges: Air").

At even higher levels of the atmosphere, gaseous atoms and molecules exist in relatively small concentrations and make up the two outer layers of the atmosphere, the mesosphere and thermosphere. Temperatures within the mesosphere generally decrease with increasing altitude, while the opposite is true for the thermosphere. High energy radiation from the sun reacts with certain gases to create a region of charged particles called the ionosphere, which is located primarily within the thermosphere. Heat is released during these reactions, making this a very warm region. Some long-distance radio communication systems depend upon the ionosphere for their transmissions. The charged particles within this region reflect the radio signals and thereby extend their effective transmission range (see Figure 2).

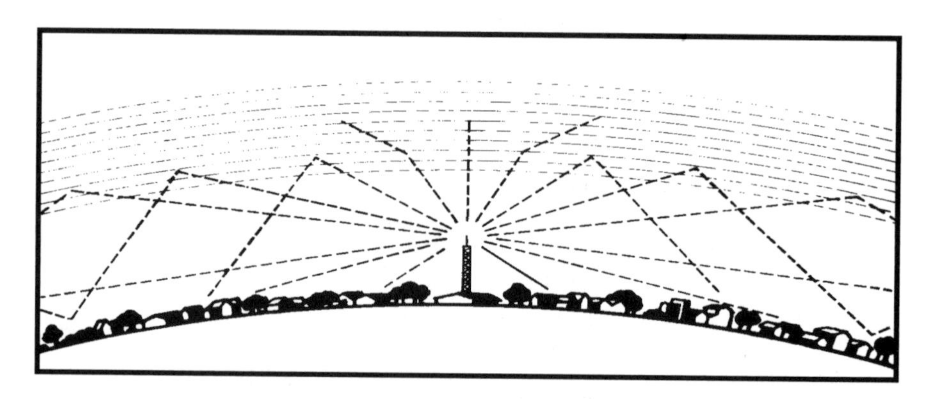

Figure 2. Earth's ionosphere reflects radio transmissions, extending their range

At the uppermost reaches of the atmosphere, high-speed particles from the sun are guided toward Earth's polar regions by Earth's magnetic field. They collide with and ionize molecules of gas, causing visible light to be emitted. The resulting brilliant displays are called the *aurora borealis* in the Northern Hemisphere and the *aurora australis* in the Southern Hemisphere.

Atmospheric Pressure

Because all molecules within the layers of the atmosphere have mass, they exert pressure on Earth and other objects within the atmosphere. Atmospheric pressure is the weight exerted by the column of air that extends to the top of the atmosphere above a specific area. Air pressure is maximum at Earth's surface and diminishes with increasing altitude, because the mass of the air above decreases. At sea level, atmospheric pressure averages about 1,013 millibars (1.04 kilograms per square centimeter). Since the force due to atmospheric pressure at any point is the same in all directions—up, down, sideways—and is the same on every part of our bodies, and since our bodies exert an equal force from within, atmospheric pressure is not felt as weight.

Atmospheric pressure can be measured by an instrument called a barometer. One type of barometer is a J-shaped glass tube containing liquid mercury with a vacuum space at the top of the tube. The height to which the mercury rises in the tube due to the weight of the atmosphere is a measure of the atmospheric pressure and is expressed in millibars (mb) (see Figure 3).

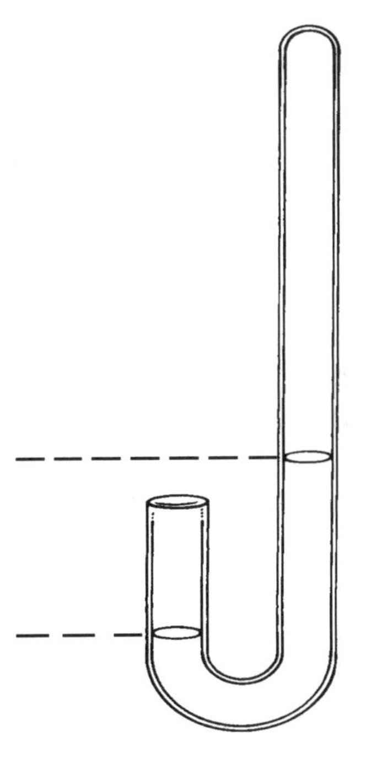

Figure 3. A mercury barometer

Composition and Evolution

Like Earth's ecological and biological systems, the atmosphere has evolved over time, changing in properties and composition. Earth's atmosphere was created from the inside out. It is believed that dissolved gases escaped from Earth's interior as the planet cooled and the crust solidified. The gravitational pull of the planet trapped the gases close to Earth.

Outgassing from Earth's interior continues today in active volcanoes and provides clues to the composition of the early atmosphere. Volcanic emissions include nitrogen (N_2), water vapor (H_2O), sulphur dioxide (SO_2), carbon dioxide (CO_2), and trace gases such as argon (Ar). Oxygen (O_2), the life-sustaining gas required by many living things, comprises over 20 percent of our present atmosphere; yet, it is conspicuously absent from the list of volcanic gases. Was oxygen present in Earth's original atmosphere? The answer may be found in the analysis of rocks that were exposed to the atmosphere very early in Earth's evolution, four billion years ago. Because oxygen is chemically active and combines readily (through oxidation) with other elements to form oxides, rust on iron and iron-containing minerals gives substantial evidence of the presence of oxygen. The fact that these early rocks show no evidence of oxidation suggests that oxygen was not present in the early atmosphere.

Examination of these early rocks and other clues provide evidence of changes in the atmosphere over time. Scientists have studied ancient air bubbles trapped in 85 million year old amber (hardened tree resin). These tiny bits of the ancient atmosphere have shown concentrations of oxygen 50 percent higher than we find in today's atmosphere. This suggests that atmospheric oxygen underwent significant fluctuations at least over the past 100 million years. What accounts for the introduction of oxygen and the differing levels of oxygen in the atmosphere from its beginning through its evolution to our present atmosphere? Some scientists theorize that very intense solar radiation in the primordial atmosphere split the oxygen from the hydrogen in water (H_2O). Others believe that the first atmospheric oxygen came from plant life. Through photosynthesis, plants consume carbon dioxide and produce oxygen. Evidence of the rapid expansion of plant life over Earth's surface is simultaneously linked in the geologic record with evidence of the appearance of oxygen in the atmosphere. This suggests that the addition of oxygen to the atmosphere occurred with the establishment of plant life. The reasons for the fluctuation in oxygen level are not completely understood.

Our Impact on the Atmosphere

The concentrations of other important gases, such as carbon dioxide, have also changed significantly over the course of the evolution of Earth's atmosphere, and the atmosphere continues to change today. A very important factor in current atmospheric evolution is that humans are now in a position to drastically alter the atmosphere's composition. Emissions from power plants, automobiles, and other human sources have an important impact on the level of pollutants in the atmosphere. Increasing amounts of certain particles and gases that trap long-wave radiation emitted by Earth may be contributing to an overall pattern of global warming (see Reading 6, "Global Warming and the Greenhouse Effect"). In addition, pollutants in the atmosphere react with one another and with naturally occurring components of the atmosphere to form potentially hazardous compounds such as acid rain (see Reading 7, "Environmental Effects of Acid Rain"). Our effect on the balance of atmospheric components translates to an impact on the entire global environment—ecology, climatic systems, and weather.

El Niño Southern Oscillation (ENSO)

SCiLINKS. THE WORLD'S A CLICK AWAY

Topic: El Niño
Go to: www.scilinks.org
Code: PESM168

A striking example of the close relationship between the ocean and the atmosphere is seen in the phenomena known as El Niño and La Niña (from the Spanish terms meaning "little boy" and "little girl"). Collectively, scientists refer to these events as the El Niño Southern Oscillation, or ENSO.

In 1997–98, the world witnessed a major ENSO event. The eastern equatorial Pacific Ocean warmed from five degrees Celsius to nine degrees Celsius above normal, resulting in atmospheric disruptions to the jet streams in both hemispheres of the globe. The end results were floods or droughts throughout much of the world's mid-latitudes. California experienced increased precipitation which led to localized flooding, mudslides, and damage to lives and property throughout much of the central and southern coasts. In contrast, the northern tier states from the Rockies to the Great Lakes enjoyed a very mild winter—golf was played all year in Minnesota!

Weather Events

In normal years, the winds tend to blow from east to west across the waters of the tropical Pacific. The easterly winds push the surface waters westward across the ocean. In turn, this causes deeper, colder waters to rise to the surface. This "upwelling" of deep ocean waters brings with it the nutrients that otherwise would remain near the bottom. The fish populations living in the upper waters are dependent on these nutrients for survival. During El Niño years, however, the winds weaken, stopping the upwelling of the colder deep water. As the ocean warms, the warmer water shifts eastward and so do the clouds and thunderstorms that produce heavy rainfall along the equator. El Niño events occur on average every three to five years.

In essence, ENSO events happen because the relationship between the ocean and the atmosphere is disturbed. Sea surface temperatures only have to increase by one degree Celsius for jet streams to bring storm activity to normally dry regions, and vice versa.

El Niño's Effects

The 1982–83 El Niño was unusually strong. In Ecuador and northern Peru, up to 250 centimeters of rain fell during a six-month period, transforming the coastal desert into a grassland dotted with lakes. Abnormal wind patterns also caused the monsoon rains to fall over the central Pacific instead of on its western edge, which led to droughts and disastrous forest fires in Indonesia and Australia.

Overall, the loss to the global economy as a result of El Niño amounted to more than $8 billion. Likewise, the record-breaking El Niño winter of 1997–1998 resulted in unusual weather in many parts of the world. In the United States, severe weather events included flooding in the Southeast, major storms in the Northeast, and flooding in California.

La Niña

By May 1998, the El Niño event was essentially over, but rapid cooling in the Eastern Tropical Pacific led to a La Niña event. La Niña is characterized by below-normal sea surface temperatures in the eastern equatorial Pacific; therefore, its effects tend to be nearly opposite those of El Niño. For example, whereas El Niño led to massive flooding in California, La Niña led to below normal precipitation in California and the Southeast.

Consider the dimensions of this climate event: a body of open ocean some 1240 km wide and 300 km from north to south across the equator, cooling by up to 10 degrees Celsius in the space of 30 days! Mother Nature demonstrated an enormous ability to adjust the thermostat of Planet Earth.

ENSO and Global Warming

Some scientists now think that these constant natural adjustments may be in part caused by the increase in the greenhouse gases that have led to increased heat energy in the global system. The El Niño, La Niña events are, in fact, linked with massive heat (and cooling) transfers. Advanced computer models, new ocean/atmosphere data buoys in the Atlantic and the Indian oceans, a system of buoys in the Pacific known as the Tropical Atmospheric Ocean Array, and more scientists trained in both oceanography and meteorology (called climatologists) will shed new light and insights on the ENSO in the future. Next to the seasons, ENSO is the strongest element affecting climate on this planet.

The Facts About Ozone

SCILINKS. THE WORLD'S A CLICK AWAY

Topic: ozone
Go to: www.scilinks.org
Code: PESM170

The Earth's ozone layer protects all life from the Sun's harmful radiation, but human activities have damaged this natural shield. Less protection from ultraviolet light will, over time, lead to higher skin cancer and cataract rates and crop damage. The United States, in cooperation with over 140 other countries, is phasing out the production of ozone-depleting substances in an effort to safeguard the ozone layer.

The Ozone Layer

The Earth's atmosphere is divided into several layers. The lowest region, the troposphere, extends from the Earth's surface up to about 10 km in altitude. Virtually all human activities occur in the troposphere. Mt. Everest, the tallest mountain on the planet, is only about nine km high. The next layer, the stratosphere, continues from 10 km to about 50 km. Most commercial airline traffic occurs in the lower part of the stratosphere.

Most atmospheric ozone is concentrated in a layer of the stratosphere about 15–30 kilometers above the Earth's surface. Normal oxygen, which we breathe, has two oxygen atoms and is colorless and odorless. Ozone is a molecule containing three oxygen atoms. It is blue in color and has a strong odor. Ozone is much less common than normal oxygen. About two million out of every10 million air molecules are normal oxygen, but only three molecules out of every 10 million are ozone.

However, even the small amount of ozone plays a key role in the atmosphere. The ozone layer absorbs a portion of the radiation from the sun, preventing it from reaching the planet's surface. Most importantly, it absorbs the portion of ultraviolet light called UVB (ultraviolet band 280–315). UVB has been linked to many harmful effects, including various types of skin cancer, and cataracts.

Ozone molecules are continually formed and destroyed in the stratosphere. The total amount, however, remains relatively stable. The concentration of the ozone layer can be thought of as a river's depth at a particular location. Although water is constantly flowing in and out, the depth remains constant.

Although ozone concentrations vary naturally with sunspots, the seasons, and latitude, these processes are well understood and predictable. Scientists have kept records spanning several decades that detail normal ozone levels during these natural cycles. Each natural reduction in ozone levels has been followed by a recovery. Recently, however, convincing scientific evidence has shown that the ozone shield is being depleted well beyond changes due to natural processes.

Ozone Depletion

For more than 50 years, chlorofluorocarbons (CFCs) were thought of as miracle substances because they are stable, nonflammable, low in toxicity, and inexpensive to produce; CFCs have been used as refrigerants, solvents, foam blowing agents, and in other smaller applications. Chlorine or bromine are known ozone-depleting substances. Other chlorine-containing compounds include methyl chloroform (a solvent) and carbon tetrachloride (an industrial chemical). Halons and methyl bromide contain bromine. The former is an extremely effective fire extinguishing agent, the latter is an effective produce and soil fumigant.

All of these compounds have atmospheric lifetimes long enough to allow them to be transported by winds into the stratosphere. Because they release chlorine or bromine when they break down, they damage the protective ozone layer. The discussion below of the ozone depletion process focuses on the long-term depletion caused by CFCs, but the basic concepts apply to all of the ozone-depleting substances including short-term depletion from volcanic eruption aerosols.

In the early 1970s, researchers began to investigate the effects of various chemicals on the ozone layer, particularly CFCs, which contain chlorine. They also examined the potential impacts of other chlorine sources. Chlorine from swimming pools, industrial plants, sea salt, and volcanoes does not reach the stratosphere. Chlorine compounds from these sources readily combine with water. Repeated measurements show that they rain out of the troposphere very quickly. In contrast, CFCs are very stable and do not dissolve in rain. Thus, there are no natural processes that remove the CFCs from the lower atmosphere; over time, winds drive the CFCs into the stratosphere.

The CFCs are so stable that only exposure to strong ultraviolet radiation breaks them down. When that happens, the CFC molecule releases atomic chlorine. One chlorine atom can destroy more than 100,000 ozone molecules. The net effect is the destruction of ozone faster than it is naturally created. To return to the analogy comparing ozone levels to a river's depth, CFCs act as a siphon, removing water faster than normal and reducing the depth of the river. Large fires and certain types of marine life produce one stable form of chlorine that does reach the stratosphere. However, numerous experiments have shown that CFCs and other widely-used chemicals produce roughly 85% of the chlorine in the stratosphere, while natural sources contribute only 15 percent.

Large volcanic eruptions can have an indirect effect on ozone levels. Although Mt. Pinatubo's 1991 eruption did not increase stratospheric chlorine concentrations, it did produce large amounts of tiny particles called aerosols (different from consumer products also known as aerosols). These aerosols increase chlorine's effectiveness at destroying ozone. The aerosols only increased depletion because of the presence of CFC-based chlorine. In effect, the aerosols increased the efficiency of the CFC siphon, lowering ozone levels more than would have occurred otherwise. Unlike long-term ozone depletion, however, this effect is short-lived. The aerosols from Mt. Pinatubo have already disappeared, but satellite, ground-based, and balloon data still show ozone depletion occurring closer to the historic trend.

One example of ozone depletion is the annual ozone "hole" over Antarctica that has occurred during the Antarctic Spring since the early 1980s. Rather than being a literal hole through the layer, the ozone hole is a large area of the stratosphere with extremely low amounts of ozone. Ozone levels fall by over 60% during the worst years.

In addition, research has shown that ozone depletion occurs over the latitudes that include North America, Europe, Asia, and much of Africa, Australia, and South America. Over the United States, ozone levels have fallen 5–10%, depending on the season. Thus, ozone depletion is a global issue and not just a problem at the South Pole.

Reductions in ozone levels will lead to higher levels of UVB reaching the Earth's surface. The sun's output of UVB does not change; rather, less ozone means less protection, and hence more UVB reaches the Earth. Studies have shown that in the Antarctic, the amount of UVB measured at the surface can double during the annual ozone hole. Another study confirmed the relationship between reduced ozone and increased UVB levels in Canada during the past several years.

Laboratory and epidemiological studies demonstrate that UVB causes non-melanoma skin cancer, plays a major role in malignant melanoma development, and is linked to cataracts.

Even with normal ozone levels all sunlight contains some UVB, so it is always important to limit exposure to the sun. However, all life forms, not just humans, are affected by increased UVB radiation. Some crops, plastics and other materials, and certain types of marine life, are particularly sensitive.

The World's Reaction

The initial concern about the ozone layer in the 1970s led to a ban on the use of CFCs as aerosol propellants in several countries, including the United States. However, production of CFCs and other ozone-depleting substances grew rapidly afterward as new uses were discovered.

Through the 1980s, other uses expanded and the world's nations became increasingly concerned that these chemicals would further harm the ozone layer. In 1985, the Vienna Convention was adopted to formalize international cooperation on this issue. Additional efforts resulted in the signing of the Montreal Protocol in 1987. The original protocol would have reduced the production of CFCs by half by 1998.

After the original Protocol was signed, new measurements showed worse damage to the ozone layer than was originally expected. In 1992, reacting to the latest scientific assessment of the ozone layer, the parties decided to completely end production of halons by the beginning of 1994 and of CFCs by the beginning of 1996 in developed countries.

Because of measures taken under the Protocol, emissions of ozone-depleting substances are already falling. Assuming continued compliance, stratospheric chlorine levels will peak in a few years and then slowly return to normal. The good news is that the natural ozone production process will heal the ozone layer in about 50 years.

The Stratospheric Protection Program

In addition to regulating the end of production of the ozone-depleting substances, the EPA implements several other programs to protect the ozone layer under Title VI of the Clean Air Act. These programs include refrigerant recycling, product labeling, banning nonessential uses of certain compounds, and reviewing substitutes.

Adapted from EPA Stratospheric Protection Division factsheet: *Ozone Science: The Facts Behind the Phaseout* http://www.epa.gov/ozone/science/sc_fact.html, June 1998.

Air Pollution and Environmental Equity

SCILINKS. THE WORLD'S A CLICK AWAY

Topic: air pollution
Go to: www.scilinks.org
Code: PESM174

A National Decision to Make Our Air Cleaner

The Clean Air Act, passed in 1963 and amended in 1970 and again in 1990 , is a decision Americans made to achieve better air quality through public policies. Based on scientific information, the Clean Air Act has built on the strengths of the scientific, environmental, and policy lessons we have learned as a nation over several generations. Understanding this relationship—between advances in scientific understanding, the decision making process, and evolving public policies—will stand your students in good stead as they make their way in the world.

The Act's overall goal is to reduce pollutants in our air by 32 billion kilograms a year—125 kilograms for every person—by the year 2005. The Clean Air Act achieves this goal by defining standards for regulating the impact of human activity on air quality. It establishes programs for monitoring human impact, for educating us about human impact, and for cleanup. It also establishes mechanisms for making sure standards are met.

Types of Pollutants

Two kinds of airborne pollutants are regulated under the Clean Air Act. The first group, called "criteria" pollutants, includes carbon monoxide, nitrogen dioxide, sulphur dioxide, ozone, lead, and particulate matter. These are discharged in relatively large quantities by a variety of sources, and they threaten human health and welfare across the country. The EPA is charged with setting national standards for each criteria pollutant, and each state takes action to ensure the national standards are met. Failure to meet national standards is called "non-attainment." Unfortunately, concentrations of human activity often lead to concentrations of pollution. Many urban areas are classified as "non-attainment areas" for at least one criteria air pollutant.

The second group of pollutants includes those that are immediately hazardous to human health. Most are associated with specific sources. Some cause cancer and some produce other health and environmental problems. This second group is large, and contains some familiar names. Benzene, for example, is a potent cancer-causing substance released chiefly through burning gasoline. Trace amounts of mercury found in coal are released when coal is burned. Mercury is also released by incinerators burning garbage and, because until 1990 it was used in latex paints to prevent mildew, when paint weathers.

Areas of non-attainment for criteria pollutants are classified according to extent of pollution. Generally speaking, the five classes range from "easy to clean up quickly" all the way to "will take a lot of work and a long time to clean up." The Clean Air Act uses these classifications to tailor cleanup requirements to the severity of the pollution, and to set realistic deadlines for achieving cleanup goals. If deadlines are missed, the Act allows more time, but usually a missed deadline in a non-attainment area means stricter requirements must then be met, such as those set for even more polluted areas.

States do most of the planning for cleaning up criteria air pollutants using a system of permits to make sure power plants, factories, automobiles, and other pollution sources meet cleanup goals. Cleaner fuels, cleaner vehicles, better maintenance programs for vehicles already on the road, and other systems are also used. If pollution should get worse, pollution control systems could be required for smaller sources of pollution.

Setting the Standards

According to the Clean Air Act, the EPA must identify categories of the major sources of these chemicals and then develop "maximum achievable control technology" standards for each category over the next 10 years. The law says these standards are to be based on the best control technologies that have been demonstrated in these industrial categories. Though EPA sets the standards, state and local air pollution agencies are primarily responsible for making sure standards are met.

The Clean Air Act favors setting standards that industry must achieve, rather than dictating equipment that industry must install. This flexibility allows industry to develop its own cost-effective means of reducing air toxics emissions and still meet the goals of the act. The Act includes unique incentives for industries to reduce their emissions early, rather than waiting for federal standards. Sources that reduce emissions by 90 percent or more before the standards go into effect will get six additional years to comply with the Act.

The Clean Air Act also establishes a program for preventing

accidental releases of air toxics from industrial plants, and for creating a Chemical Safety Board to investigate accidental releases that occur. Between 1982 and 1986, accidental releases of toxic chemicals in the United States caused 309 deaths, 11,341 injuries, and the evacuation of 464,677 people from homes and jobs.

The Clean Air Act establishes "enforcement" methods that can be used to make polluters obey laws and regulations. Enforcement methods include citations (like traffic tickets), fines, and even jail terms. The knowing violation of almost every requirement is now a felony offense. EPA and state and local governments are responsible for enforcing the Clean Air Act but, if they don't, members of the public can also sue EPA or the states to get action. Citizens also can sue violators apart from any action taken by EPA or state or local governments.

Environmental Equity

EPA is required by law to take environmental equity into account in all its policy decisions. Environmental equity, also called environmental justice, describes the perception of fairness in how environmental quality is distributed across groups of people who have different characteristics. The environmental impact of a human activity, for example, is evaluated by EPA to determine how benefits, risks, and harm are distributed among people categorized according to gender, age, ethnicity, place of residence, occupation, income, etc.. Currently, environmental equity refers more specifically to how health risks are distributed across such groups. In the context of air quality, this generally refers to how exposed certain groups of people are to airborne toxic substances.

Early Beginnings

The first studies documenting the relationship between the geographic distribution of environmental pollution and the geographic distribution of minority populations were published in the early 1970s. In its 1971 *Report to the President of the United States*, the Council on Environmental Quality acknowledged that racial discrimination adversely affects the ability of urban poor to control environmental quality where they live.

Environmental equity did not become a national issue, however, until 1982, when a predominatly African-American community in Warren County, North Carolina, focused public attention on a government proposal to dump PCBs (polychlorinated biphenyls) in their community. Since then, it has become generally accepted that minorities have the potential to be disproportionately exposed to toxic substances in their residential environments. In 1983, the U.S. General Accounting Office found that three out of four commercial hazardous waste sites in the southeastern United States are located in predominantly non-white communities.

Another study of this relationship was conducted in Detroit, Michigan, in the late 1980s. After comparing the influence of both income and ethnicity on waste facility distribution, investigators published documentation that ethnic minority residents were four times more likely to live within a mile of a commercial hazardous waste facility, and that ethnicity was a better predictor of resident proximity to such facilities than was income.

Future Goals

As a result of these and many other efforts, by the early 1990s environmental equity had moved from grassroots social movement through formal academic research into federal environmental policy. In 1994, Executive Order 12898 was issued by President Bill Clinton requiring all federal agencies to make achieving environmental equity an integral part of their mission. The EPA has established outreach strategies, internal offices, task forces, grant programs, advisory councils, and oversight committees designed specifically to address equity issues in environmental policy. EPA is also required by Executive Order 12898 to address environmental equity in all its education materials, and this book is one among many efforts to fulfill that obligation.

Adapted from *Investigating Air*, National Science Teachers Association, 1999.

Weather and the Redistribution of Thermal Energy

Topic: weather and energy
Go to: www.scilinks.org
Code: PESM178

When we describe weather, we typically speak in terms of temperature, humidity, wind, and the presence or absence of precipitation. Weather, however, is not a set of random acts of nature, but a response to the unequal heating of Earth's atmosphere. Imbalances in rates of heating and cooling from one place to another within the atmosphere create temperature gradients. In response to these gradients, the atmosphere circulates and thermal energy is redistributed. While there are a number of complicating factors within the redistribution equation, it is important for students to understand this basic premise.

Heat versus Temperature

It is important to note the distinction between *temperature* and *heat*. The most common descriptions of weather generally relate to temperature: "How cold is it outside?" or "Is it going to be hot today?" However, the *redistribution of heat* is the driving force behind weather. What is the difference between temperature and heat? Temperature is defined as the *average* kinetic energy, or energy of motion, per atom or molecule of a particular substance. The greater the kinetic energy of the atoms or molecules, the higher the temperature. Heat, on the other hand, is defined as the *total* kinetic energy of all of the atoms or molecules composing a given amount of a substance.

The distinction between heat and temperature can be appreciated through the following example. Compare a pot of water at 90° C, close to the boiling point of water, with a bathtub filled with water at 40° C. The water in the pot has a higher *temperature* than the water in the tub —i.e. the average kinetic energy of the water molecules in the pot is higher than that in the tub. However, the water in the tub has more *heat* because the tub contains so many more molecules of water. Heat, remember, is the *total amount* of kinetic energy of *all* of the atoms or molecules composing a substance.

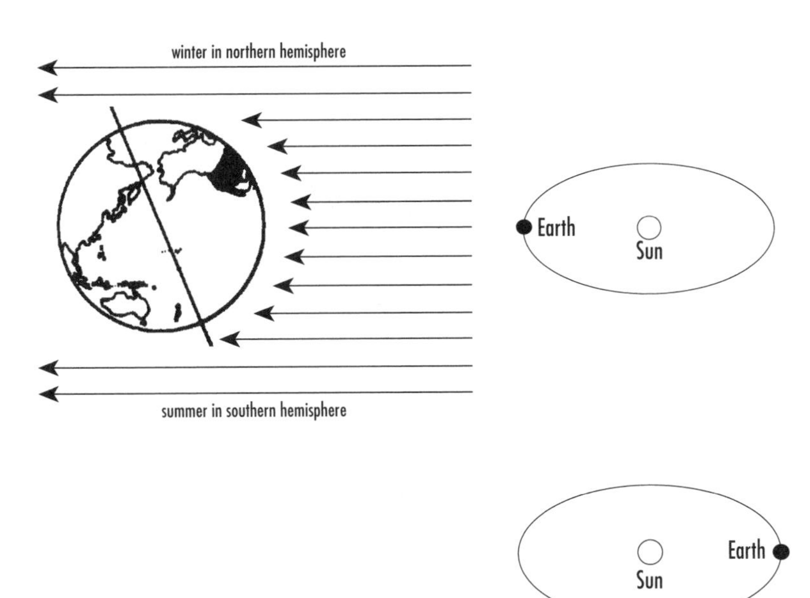

Figure 1. The amount of radiant energy absorbed on Earth depends on the number of daylight hours and on the incoming angle of solar rays. Compare the incoming angle of solar rays during winter and summer in the different hemispheres.

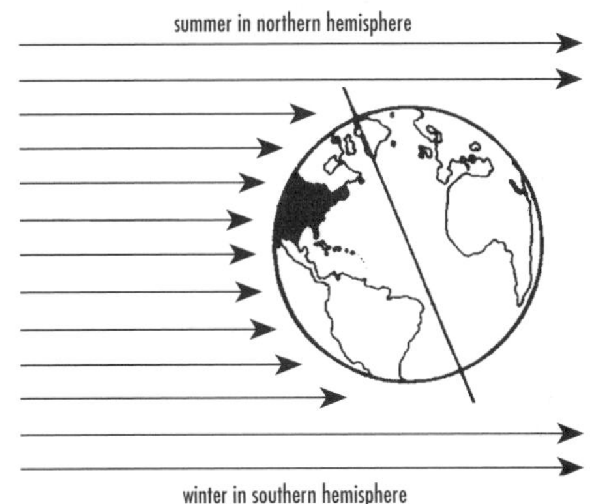

Heating the Earth

Earth's atmosphere is heated by solar radiation and by re-radiation of solar energy reflected from Earth's surface. For all practical purposes we can say that Earth receives nearly a constant rate of radiant energy from the sun. This energy, however, is not uniformly distributed throughout the planet. The 23.5 degree tilt in Earth's axis causes maximum intensities of solar radiation to strike the northern hemisphere during the middle months of the year and the southern hemisphere during the beginning and ending of the year (see Figure 1).

In addition, the curvature of Earth's surface affects the distribution of solar energy. At the equator, the sun's rays fall most nearly perpendicular. This transmits the highest amount of solar radiation because those rays strike a smaller surface area than do rays striking near the poles. This concentrating effect means that the amount of energy per square unit of surface area is greater near the Equator than near the poles. (see Figure 2; this is demonstrated in Activity 6, "Why is it Hotter at the Equator than at the Poles?")

The atmosphere is transparent to most incident solar radiation. However, some radiation is absorbed, scattered, or reflected by the atmosphere, depending on its wavelength. Radiation of some wavelengths is

Figure 2. Earth's shape affects how its surface heats. More heat is transferred where the sun's rays strike Earth directly (near the equator) than where they strike Earth slanted (near the poles). Slanted rays transfer less heat largely because they are dispersed over a much larger area than are direct rays. This principle is demonstrated by measuring the area covered by a flashlight's beam on perpendicular and slanted surfaces.

absorbed by water vapor, ozone, and dust particles; other wavelengths are scattered by air molecules; still others are reflected by clouds. A large portion of the total solar radiation reaching Earth passes through the atmosphere and reaches the ground, where it either is reflected or absorbed. Some land materials—e.g. rocks, snow, and sand—readily reflect most of the sun's radiations. In contrast, bodies of water absorb, rather than reflect, most of the radiations they receive.

Heat Transfer Within the Atmosphere

Overall, Earth's *atmosphere* transmits, scatters, and reflects more radiant energy from the sun than it absorbs. Earth's *surface*, on the other hand, absorbs more solar energy, on average, than it reflects. On knowing only these facts, one might expect the atmosphere to be cooling while the surface heats. However, this is not the case because that imbalance is counteracted by the transfer of heat energy from the surface back to the atmosphere.

This transfer of heat occurs primarily through two different but interactive mechanisms: sensible heating and latent heating. Sensible heating involves the processes of conduction and convection. It accounts for about 23 percent of the overall heat energy transferred into the atmosphere from Earth's surface. Latent heating involves the transfer of heat as a consequence of changes in phase of water. This kind of heating accounts for about 77 percent of the heat transferred from Earth's surface to the atmosphere.

One mechanism of sensible heating, conduction, involves the transfer of heat energy from a warmer object to a cooler one through direct contact. Conduction is the principle underlying why the handle of a fireplace poker becomes hot when just the tip is left in a fire. Heat energy is transferred from the fire to the tip of the poker, and the metal in the poker transfers (i.e. conducts) the heat energy from one end to the other. Within the atmosphere, conduction is significant only in a very thin layer of air that is in immediate contact with Earth's surface.

Convection, on the other hand, is the process of heat distribution within a fluid (such as air), achieved through movement of the fluid itself. Convection is an important process in atmospheric heating. It results from density differences between

Figure 3.
Heat transfer through convection

parcels of air with differing temperatures. The process of atmospheric convection begins when a parcel of air near Earth's surface is warmed. Warmer air is less dense than cooler air, thus it rises away from the surface. As it rises, it is replaced by cooler air underneath. That cooler air may, in turn, be warmed by the surface, become less dense and rise, repeating the process. As the warm air rises, it expands and cools, becoming more dense and sinking. Convection currents or cells are established through this process of heating and cooling. This circulation transports heat energy into the atmosphere (see Figure 3). The principle of convection is used in home heating systems, with heaters or hot air vents usually placed at floor level rather than near the ceiling.

The convection process is facilitated by changes in air pressure. As warm air rises, its pressure decreases, causing the air to expand and cool. As cool air sinks, its pressure increases and it is compressed and warmed. The expansional cooling and compressional warming mechanisms are important aspects of the atmospheric convection process.

As previously mentioned, latent heating is a major mechanism for atmospheric heating, much more so than is sensible heating. Latent heat is the heat energy released or absorbed when a substance changes phase: from solid to liquid, liquid to gas, gas to liquid, etc. The latent heat associated with changes in phase of water is described in Figure 4, which illustrates how energy is absorbed and released as one gram of ice at –10° C is transformed into one gram of water vapor at 110° C. As heat energy is applied to the ice, its temperature increases until it

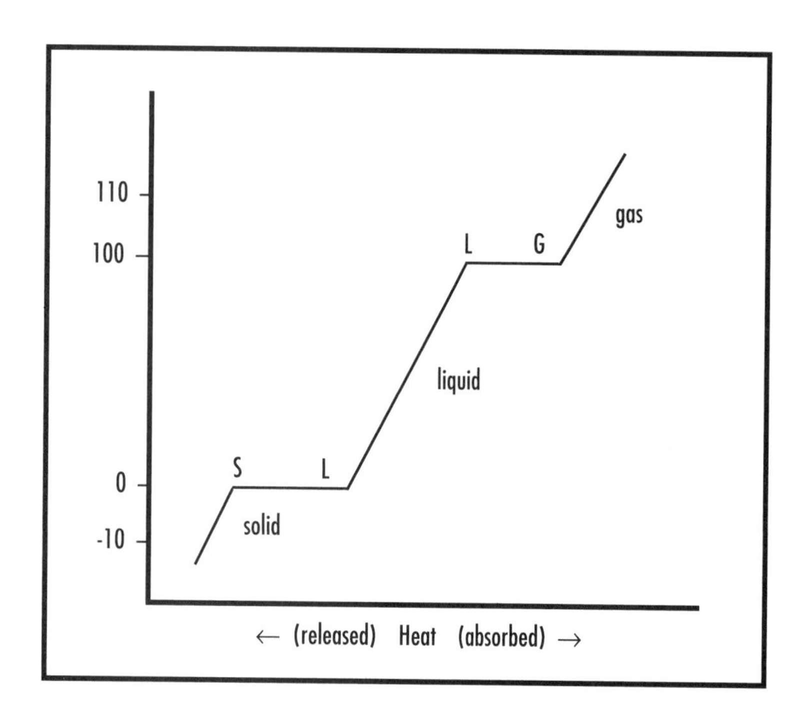

Figure 4. Changes of phase for water

reaches 0° C, the melting point of water. An additional amount of heat energy must be absorbed by the water to change its phase from solid to liquid (between points S and L on the graph in Figure 4). The amount of heat energy required for this transformation is called the *latent heat of fusion*. During the addition of the latent heat of fusion, the temperature of the water remains constant until all of the solid ice is melted into liquid water.

As more heat energy is added to the gram of (now liquid) water, its temperature increases steadily until it reaches 100° C, water's boiling point. Just as with the transition from solid to liquid, an additional amount of heat energy must be absorbed by the water to change its phase from liquid to vapor. This amount of heat energy is called the *latent heat of vaporization*. As during melting, the water's temperature remains constant until all the liquid is evaporated to water vapor, a gas.

As more heat energy is added to the gram of water vapor, the temperature of the gas increases from 100° C to 110° C.

If the process were reversed, the same amount of heat energy that was absorbed when the water was warmed would be released as the water cooled. Heat energy is released when the gas cools, when the gas condenses into liquid water, when the liquid cools, when the liquid freezes into a solid, and when the ice cools further.

Latent heating of Earth's atmosphere occurs as heat energy from the sun causes water on Earth's surface to evaporate into the overlying air. When liquid water from the surface evaporates, the latent heat of vaporization is stored in the water vapor. As air containing the water vapor is warmed by Earth's surface (through conduction and convection), it rises, carrying the water vapor with it. As the air rises, it expands and cools, and when the air becomes saturated some of the water vapor condenses into water droplets. This change of phase releases the stored latent heat of vaporization into the atmosphere, where it warms the surrounding air further and causes it to rise more. As the rising air expands and cools, additional water vapor condenses, additional heat is released, and the process continues. It is important to note that the process of latent heating builds on the process of sensible heating and allows considerable heat energy to be transferred from Earth's surface into its atmosphere.

Local Effects of Heat Transfer

Because heating of Earth's surface is uneven, the transfer of heat energy to the atmosphere varies on a *local* as well as *global* scale. Variations in atmospheric heating, in turn, results in the development of temperature gradients within the atmosphere. As is the case in other situations, there is a tendency toward evening out this heat distribution. Air moves and circulates (e.g. in convection cells) as a result. This motion causes winds. From the slightest breeze to a raging hurricane, temperature gradients are responsible for producing wind. It is important to remember, however, that winds can be created on a local as well as a global level.

An example of a local wind system caused by temperature gradients is the sea breeze/land breeze system that develops at a seashore or lake shore (see Figure 5). Bodies of water change temperature more slowly than do land masses. Thus, a sea or lake shore heats faster than does the nearby water. During the day, the air over land tends to be warmer and has a lower density than does the air over the lake or ocean. The warm air above the land rises and the cooler air above the water moves in under the warm air to replace it; a sea breeze is established. At night, the land cools faster than the ocean, with the air above the land becoming cooler and more dense. The relatively warm air over the water rises and a land breeze results as the movement of cool air towards the water occurs.

Figure 5. Land and sea breezes result from differential heating and cooling of land and water

Global patterns of air motion likewise develop from temperature gradients within the atmosphere. The distribution of heat from the warmer regions of Earth's surface throughout the rest of the globe is largely achieved by global winds in combination with deep ocean currents. Deep water ocean currents slowly circulate cold water along the ocean bottom in a general direction from the poles toward the Equator. Toward the surface warm water moves from near the Equator toward the poles to complete the convection cell. Likewise, warm and moist air from equatorial regions rises, while cool and dry air from the poles sinks and moves underneath the rising warm air. This produces a global wind pattern. (The pattern is not strictly a circular cell because Earth's rotation deflects winds to the right of their direction of motion in the northern hemisphere and to the left of their direction of motion in the southern hemisphere. This deflection is called the Coriolis Effect (see Reading 9, "The Inner Workings of Severe Weather").

Air Masses and Fronts

A region or body of air that has consistent temperature and moisture content throughout is called an air mass. When two or more air masses with significantly different properties meet, they do not readily mix. The two masses interact along a boundary called a frontal zone or front. There are several types of fronts, including:

Cold front: A cold air mass advances against a warm air mass, forcing the warm air upward. Clouds, precipitation, and sometimes severe weather result during the passage of a cold front. Cooler, dryer air moves into an area after the passage of a cold front.

Warm front: A warm air mass advances against a cooler air mass, riding up over the cooler air in front of it. Clouds and precipitation in the form of rain, snow, sleet or freezing rain can result during the passage of a warm front. Warmer moist air moves into an area after the passage of a warm front.

Stationary front: A condition where neither the cold air mass nor the warm air mass can advance against the other. The interaction of warm and cool air along the front is responsible for rain, thunderstorms, and snow.

Frontal zones are responsible for the formation of much of the cloudiness, rain, and snow that occurs over the United States, especially in winter. At a frontal zone, the warmer moist air rises and cools, while cooler, dryer air sinks and warms. As the warm air mass rises, it expands into the lower pressure environment aloft and cools. As it cools, its capacity to hold water vapor decreases, and condensation of water vapor releases latent heat, causing further lifting of the air mass. Winds develop as the warmer rising air results in lower air pressure near Earth's surface and cooler air moves into the low pressure area. A cloud begins to form when the air becomes saturated—i.e. when the dew point temperature is reached (see Reading 8, "Weather's Central Actor: Water").

Although extremely complex, both global and local weather systems are based in relatively simple processes of heat transfer. Fronts, wind patterns, and convection cells each result from the unequal accumulation of thermal energy over Earth's surface and the mechanisms that exist for its redistribution. Understanding these basic processes forms the foundation for understanding the weather around us.

Global Warming and the Greenhouse Effect

SCILINKS. THE WORLD'S A CLICK AWAY
Topic: greenhouse effect
Go to: www.scilinks.org
Code: PESM186

The "greenhouse effect" has caught the imagination of the general population in the last two decades. What's more, the respected, generally conservative scientific establishment has become associated with dire predictions of future climate changes the greenhouse effect may cause. But how much do we actually know about the greenhouse effect? Can we really establish how much the climate will change, and when? The questions below will help outline what is currently believed about the greenhouse effect.

What Is the Greenhouse Effect, and Is It Affecting Our Climate?

The greenhouse effect is unquestionably real, and is essential for life on Earth. It is the result of heat absorption by certain gases in the atmosphere (called greenhouse gases because they trap heat) and re-radiation downward of a part of that heat. Water vapor is the most important greenhouse gas, followed by carbon dioxide and other trace gases. Without the greenhouse effect, the average temperature of Earth's surface would be about -18° C (0° F) instead of its present 14° C (57° F).

Are Greenhouse Gases Increasing?

Human activity has been increasing the concentration of greenhouse gases in the atmosphere (mostly carbon dioxide from combustion of coal, oil, and gas; plus a few other trace gases). There is no scientific debate on this point. At rates of increase observed over the past few decades, the concentration of carbon dioxide will be double that of preindustrial levels in about 2050.

Is the Climate Warming?

Global surface temperatures have increased approximately 0.3 to 0.6°C (1° F) since the late 19th century, and about 0.5° F over the past 40 years (the period with the most credible data). The warming has not been globally uniform. Some areas (including parts of the southeastern United States) have cooled. The recent warmth has been greatest over North America and Eurasia between 40 and 70° north latitude. Warming, assisted by the record El Niño of 1997–1998, has continued right up to the present.

An enhanced greenhouse effect is expected to cause cooling in higher parts of the atmosphere because the increased "blanketing" effect in the lower atmosphere holds in more heat. Cooling of the lower stratosphere (at about 9,500 to 11,500 m. or 30–35,000 ft.) since 1979 is shown by both satellite Microwave Sounding Unit and radiosonde data (A radiosonde is a ballone-borne package of instruments.)

There has been a general, but not global, tendency toward reduced diurnal temperature range (the difference between high and low daily temperatures) over more than 40% of the global land mass since the middle of the 20th century. Cloud cover has increased in many of the areas with reduced diurnal temperature range.

Relatively cool surface and tropospheric temperatures, and a relatively warmer lower stratosphere, were observed in 1992 and 1993, following the 1991 eruption of Mt. Pinatubo. The warming reappeared in 1994. A dramatic global warming, at least partly associated with the record El Niño, began in mid-1997 and continues as this is written. This warming episode is reflected from the surface to the top of the troposphere.

Indirect indicators of warming such as borehole temperatures, snow cover, and glacier recession data, are in substantial agreement with the more direct indicators of recent warmth.

No long-term trend can be identified in global or hemispheric sea-ice cover since 1973 when satellite measurements began, although Northern Hemisphere sea ice coverage has been generally below average since the early 1990s.

Are El Niños Related to Global Warming?

El Niños are not caused by global warming. Clear evidence exists from a variety of sources (including archaeological studies) that El Niños have been present for hundreds, and some indicators suggest maybe millions, of years. However, it has been hypothesized that warmer global sea surface temperatures can enhance the El Niño phenomenon. It is also true that El Niños have been more frequent and intense in recent decades.

Is the Atmospheric/Oceanic Circulation Changing?

A rather abrupt change in the El Niño Southern Oscillation (ENSO) behavior occurred around 1976–77 and the pattern has persisted. There have been relatively more frequent El Niño episodes, with only rare excursions into the other extreme (cold phase, or La Niña episodes) of the phenomenon. This behavior, and especially the recurring El Niño events since 1990, is highly unusual in the last 120 years (the period of instrumental record). Changes in precipitation over the tropical Pacific are related to this change in the ENSO, which has also affected the pattern and magnitude of surface temperatures.

Is the Climate Becoming More Variable or Extreme?

On a global scale there is little evidence of sustained trends in climate variability or extremes. This perhaps reflects inadequate data and a dearth of analyses. However, on regional scales, there is clear evidence of changes in variability or extremes.

In areas where a drought usually accompanies an El Niño event, droughts have been more frequent in recent years. Other than these areas and the few areas with longer term trends to lower rainfall, little evidence is available of changes in drought frequency or intensity. In some areas there is evidence of increases in the intensity of extreme rainfall events, but no clear global pattern has emerged. Despite the occurrence in recent years of several regional-scale extreme floods there is no evidence of wide-spread changes in flood frequency. This may reflect the dearth of studies, definition problems, and/or difficulties in distinguishing the results of land use changes from meteorological effects.

There is some evidence of recent (since 1988) increases in extreme extratropical cyclones over the North Atlantic. Intense tropical cyclone activity in the Atlantic appears to have decreased over the past few decades. Elsewhere, changes in observing systems confound the detection of trends in the intensity or frequency of extreme synoptic systems. There has been a clear trend toward fewer extremely low minimum temperatures in several widely-separated areas in recent decades. Widespread significant changes in extreme high temperature events have not been observed. There is some indication of a decrease in day-to-day temperature variability in recent decades.

How Important Are These Changes in a Longer-Term Context?

For the Northern Hemisphere summer temperature, recent decades appear to be the warmest since at least about 1400 AD, and the warming since the late 19th century is unprecedented over the last 600 years. Older data are insufficient to provide reliable hemispheric temperature estimates. Ice core data suggest that the 20th century has been warm in many parts of the globe, but also that the significance of the warming varies geographically, when viewed in the context of climate variations of the last millennium.

Large and rapid climatic changes affecting the atmospheric and oceanic circulation and temperature, and the hydrological cycle, occurred during the last ice age and during the transition towards the present Holocene period (which began about 10,000 years ago). Based on the incomplete evidence available, the projected change of 1–3.5° C (2–7° F) over the next century would be unprecedented in comparison with the best available records from the last several thousand years.

Is Sea Level Rising?

Global mean sea level has been rising at an average rate of 1–2 mm/year over the past 100 years, which is significantly larger than the rate averaged over the last thousand years.

Can the Observed Changes Be Explained by Natural Variability, Including Changes in Solar Output?

Some changes, particularly part of the pre-1960 temperature record, are well correlated with solar output, but the more recent warm era is not well correlated. The exact magnitude of purely natural global mean temperature variance is not known precisely, but model experiments excluding solar variation indicate that it is probably less than the variability observed during this century

Adapated from National Oceanic and Atmospheric Administration Factsheet *Global Warming Frequently Asked Questions*, July 20, 1998 http://www.ncdc.noaa.gov/ol/climate/globalwarming.html and *EPA Journal*, January/February 1989, pp. 4-7.

Environmental Effects of Acid Rain

SCILINKS.
THE WORLD'S A CLICK AWAY

Topic: acid rain
Go to: www.scilinks.org
Code: PESM190

Air Pollution Creates Acid Rain

Scientists have discovered that air pollution from the burning of fossil fuels is the major cause of acid rain. Acidic deposition, or acid rain as it is commonly known, occurs when emissions of sulfur dioxide (SO_2) and oxides of nitrogen (NOx) react in the atmosphere with water, oxygen, and oxidants to form various acidic compounds. This mixture forms a mild solution of sulfuric acid and nitric acid. Sunlight increases the rate of most of these reactions.

These compounds then fall to the earth in either wet form (such as rain, snow, and fog) or dry form (such as gas and particles). About half of the acidity in the atmosphere falls back to Earth through dry deposition as gases and dry particles. The wind blows these acidic particles and gases onto buildings, cars, homes, and trees. In some instances, these gases and particles can eat away the things on which they settle. Dry deposited gases and particles are sometimes washed from trees and other surfaces by rainstorms. When that happens, the runoff water adds those acids to the acid rain, making the combination more acidic than the falling rain alone. The combination of acid rain plus dry deposited acid is called acid deposition. Prevailing winds transport the compounds, sometimes hundreds of miles, across state and national borders.

Electric utility plants account for about 70% of annual SO_2 emissions and 30% of NO_x emissions in the United States. Mobile sources (tranportation) also contribute significantly to NO_x emissions. Overall, more than 20 million tons of SO_2 and NO_x are emitted into the atmosphere each year.

Acid rain causes acidification of lakes and streams and contributes to the damage of trees at high altitudes. In addition, acid rain accelerates the decay of building materials and paints, including irreplaceable buildings, statues, and sculptures that are part of our nation's cultural heritage. Before falling to the earth, SO_2 and NO_x gases—and their particulate matter derivatives sulfates and nitrates—contribute to visibility degradation and impact public health.

Implementation of EPA's Acid Rain Program under the 1990 Clean Air Act Amendments will confer significant benefits.

By reducing SO_2 and NO_x, many acidified lakes and streams will improve substantially so that they can once again support fish life. Visibility will improve, allowing for increased enjoyment of scenic vistas across our country, particularly in National Parks. Stress to our forests along mountain ridges from Maine to Georgia will be reduced. Deterioration of our historic buildings and monuments will be slowed. Finally, reductions in SO_2 and NO_x will reduce sulfates, nitrates, and ground level ozone (smog), leading to improvements in public health.

Surface Waters

Acid rain primarily affects sensitive bodies of water, that is, those that rest atop soil with a limited ability to neutralize acidic compounds (called "buffering capacity"). Many lakes and streams examined in a National Surface Water Survey (NSWS) suffer from chronic acidity, a condition in which water has a constant low pH level. The survey investigated the effects of acidic deposition in more than 1,000 lakes larger than 10 acres and in thousands of miles of streams believed to be sensitive to acidification. Of the lakes and streams surveyed in the NSWS, acid rain has been determined to cause acidity in 75% of the acidic lakes and about 50% of the acidic streams. Several regions in the United States were identified as containing many of the surface waters sensitive to acidification. They include, but are not limited to, the Adirondacks, the mid-Appalachian highlands, the upper Midwest and the high elevation West.

Acid rain control will produce significant benefits in terms of lowered surface water acidity. If acidic deposition levels were to remain constant over the next 50 years (the time frame used for projection models), the acidification rate of lakes in the Adirondacks that are larger than 10 acres would rise by 50% or more. Scientists predict, however, that the decrease in SO_2 emissions required by the Acid Rain Program will significantly reduce acidification due to atmospheric sulfur. Without the reductions in SO_2 emissions, the proportions of acidic aquatic systems in sensitive ecosystems would remain high or dramatically worsen.

The impact of nitrogen on surface waters is also critical. Nitrogen plays a significant role in episodic acidification and new research recognizes the importance of nitrogen in long-term chronic acidification as well. Furthermore, the adverse impact of atmospheric nitrogen deposition on estuaries and other large water bodies may be significant. For example, 30 to 40% of the nitrogen in the Chesapeake Bay comes from atmospheric deposition. Nitrogen is an important factor in causing eutrophication (oxygen depletion) of water bodies.

Forests

Acid rain has been implicated in contributing to forest degradation, especially in high-elevation spruce trees that populate the ridges of the Appalachian Mountains from Maine to Georgia, including national park areas such as the Shenandoah and Great Smoky Mountain national parks. There also is a concern about the impact of acid rain on forest soils. There is good reason to believe that long-term changes in the chemistry of some sensitive soils may have already occurred as a result of acid rain. As acid rain moves through the soils, it can strip away vital plant nutrients through chemical reactions, thus posing a potential threat to future forest productivity.

Visibility

Sulfur dioxide emissions lead to the formation of sulfate particles in the atmosphere. Sulfate particles account for more than 50% of the visibility reduction in the eastern part of the United States, affecting our enjoyment of national parks, such as the Shenandoah and the Great Smoky Mountains. The Acid Rain Program is expected to improve the visual range in the eastern United States by 30%. Based on a study of the value national park visitors place on visibility, the visual range improvements expected at national parks of the eastern United States will be worth a billion dollars by the year 2010.

Materials

Acid rain and the dry deposition of acidic particles are known to contribute to the corrosion of metals and deterioration of stone and paint on buildings, cultural objects, and cars. The corrosion seriously depreciates the objects' value to society. Dry deposition of acidic compounds can also dirty buildings and other structures, leading to increased maintenance costs. To reduce damage to automotive paint caused by acid rain and acidic dry deposition, some manufacturers use acid-resistant paints, at an average cost of $5 for each new vehicle (or a total of $61 million per year for all new cars and trucks sold in the United States) The Acid Rain Program will reduce damage to materials by limiting SO_2

emissions. The benefits of the Acid Rain Program are measured, in part, by the costs now paid to repair or prevent damage—the costs of repairing buildings, using acid-resistant paints on new vehicles, plus the value that society places on the details of a statue lost forever to acid rain.

Health

Based on health concerns, SO_2 historically has been regulated under the Clean Air Act. Sulfur dioxide interacts with oxygen in the atmosphere to form sulfate aerosols, which may be transported long distances through the air. Most sulfate aerosols are particles that can be inhaled. In the eastern United States, sulfate aerosols make up about 25% of these particles. According to recent studies at Harvard and New York Universities, higher levels of sulfate aerosols are associated with increased sickness and death from lung disorders, such as asthma and bronchitis. When fully implemented by the year 2010, the public health benefits of the Acid Rain Program will be significant, due to decreased mortality, hospital admissions, and emergency room visits.

Decreases in nitrogen oxide emissions are also expected to have a beneficial impact on health effects by reducing the nitrate component of inhalable particulates and reducing the nitrogen oxides available to react with volatile organic compounds and form ozone. Ozone impacts on human health include a number of morbidity and mortality risks associated with lung disorders.

Clean Air for Better Life

By reducing SO_2 emissions by such a significant amount, the Clean Air Act promises to confer numerous benefits on the nation. Scientists project that the 10 million-ton reduction in SO_2 emissions should significantly decrease or slow down the acidification of water bodies and will reduce stress to forests. In addition, visibility will be significantly improved due to the reductions, and the lifespan of building materials and structures of cultural importance should lengthen. Finally, the reductions in emissions will help to protect public health.

Adapted from *EPA: Environmental Effects of Acid Rain*, http://www.epa.gov/acidrain/effects/envben.html, April 1999.

Weather's Central Actor: Water

Water is a special substance, with properties very different from those of other compounds. It is the only substance on Earth that occurs naturally in all three states—solid, liquid, and gas. Liquid water covers nearly two-thirds of Earth's surface and water in one form or another affects almost all living and non-living things. As described in Reading 5, water may change phase in response to changing conditions of temperature and pressure. Although the state and location of a molecule of water generally changes over time, the total quantity of water in all forms in Earth's hydrosphere is constant.

Hydrologic Cycle

The hydrologic or water cycle describes how this fixed amount of water moves through the environment (see Figure 1). Water is evaporated or transpired (released by plants) into the atmosphere, where it eventually condenses and returns to Earth's surface as precipitation. Here it may be temporarily stored in glaciers, lakes, underground reservoirs, or living things before returning by rivers to the oceans, or again be transpired or evaporated directly back into the atmosphere. At any one time,

Figure 1. The water cycle

about 97 percent of Earth's water is in the ocean, 2.2 percent is frozen in ice caps and glaciers, and a mere 0.001 percent is contained in the atmosphere. The time that any given water molecule spends in one form or location may be fairly long, but the atmospheric part of the hydrologic cycle is especially dynamic.

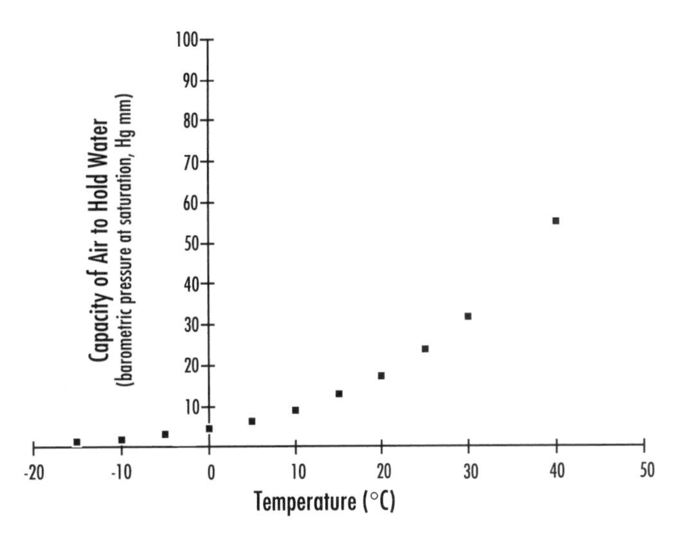

Figure 2. The atmosphere's capacity to hold water is a function of the air temperature. Relatively hot air can hold a considerably larger amount of water than can cooler air.

The relatively small amount of water in the atmosphere is essential in creating Earth's weather systems and climate. Atmospheric heating, humidity, cloud formation, and precipitation are all directly related to the amount of water vapor (gaseous water) in the atmosphere.

The amount of water vapor in a given region of the atmosphere depends on the temperature of the air and the source of the water. The capacity of air to hold water vapor increases exponentially with the temperature of the air, so that warm air can hold much more water vapor than cool air (see Figure 2). A measure of moisture content is relative humidity. It is defined as the amount of water vapor in the air relative to the maximum amount the air could hold at that temperature. The amount is expressed as a percentage and termed relative humidity because the same percentage for two different temperatures represents different actual amounts of moisture in the air. Relative humidity is generally spoken of in terms of perceived humidity or comfort level.

When air holds all the water vapor it can, the air is said to be saturated, and the relative humidity is reported as 100 percent. If warm, unsaturated air is cooled, it will eventually reach a temperature where the amount of water vapor already in the air will equal the total amount that the air can hold. At this temperature, called the dew point, the air becomes saturated and the water vapor will begin to condense to form water droplets, clouds, fog, or dew. The more water there is in unsaturated air, the less the air has to be cooled for condensation to occur. For this reason, the dew point can also be used as a measure of the actual amount of water vapor in the air. If the dew point is very close to the air temperature, then the air is said to be humid.

Water's Chemistry

The behavior of water in all states—solid, liquid, and gas—is largely attributable to its molecular structure. The water molecule is made up of one oxygen atom and two hydrogen atoms. The arrangement of the atoms within a water molecule results in polarity, meaning that the water molecules have slightly positive and negative charges on opposite sides. The hydrogen side carries a slightly positive charge while the oxygen side carries a slightly negative charge (see Figure 3).

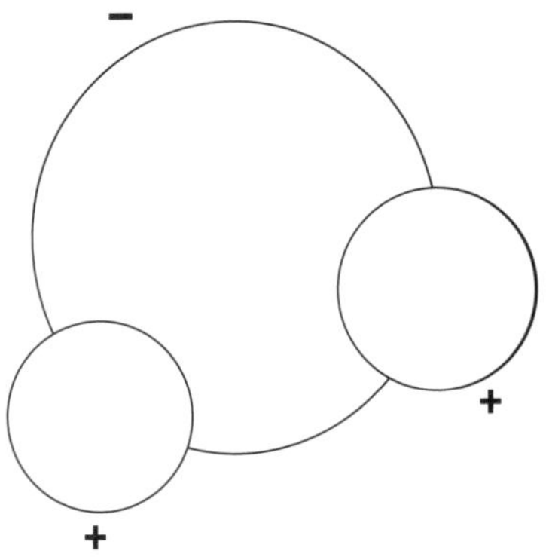

This polar configuration of the water molecule is the most stable arrangement of the oxygen and hydrogen atoms. The arrangement allows each hydrogen atom to share a pair of electrons with the oxygen atom through a covalent (shared electrons) bond.

While covalent bonds exist within water molecules, a cohesive force called hydrogen bonding acts between water molecules. Hydrogen bonding results from the attraction between the positively charged side of one molecule and the negatively charged side of another. Hydrogen bonds are largely responsible for properties of liquid water such as its dissolving power, surface tension, capillary action, and droplet formation.

Figure 3. Schematic representation of a water molecule. Although neutral overall, the two hydrogen atoms each carry a slight positive charge while the one oxygen atom carries a slight negative charge.

Further, water must absorb or release a significant amount of thermal energy before an actual change in its temperature is observed. This is also a result of the numerous hydrogen bonds that occur between water molecules. These hydrogen bonds must be broken before increased molecular motion can cause a temperature increase to occur. This property is referred to as specific heat. The specific heat of a material indicates the amount of energy required to change the temperature of the material. The greater the specific heat of a substance, the more energy absorption required to increase its temperature. Similarly, the greater the specific heat of a substance, the greater the amount of energy will be released when its temperature is lowered.

The large specific heat of liquid water and the large quantity of water in the ocean allow the ocean to absorb and store a large percentage of the radiant energy (heat) that Earth receives from the sun. This results in slow, incremental changes in ocean temperatures throughout the year as compared to the relatively quick temperature changes of the continents. The ocean is generally warmer than the continents in winter and cooler then the continents in summer. Because water absorbs and retains so much heat, the ocean tends to moderate Earth's air temperature and climate, especially along coastal regions, where it buffers temperature fluctuations.

Water also has a large latent heat (the amount of energy associated with changes of state; latent heat is described in Reading 5, "Weather and the Redistribution of Thermal Energy"). This property of water results from the hydrogen bonding between water molecules. In the solid (ice) state, water molecules are hydrogen bonded to each other in a regular crystal lattice. In liquid water, molecules of water line up in chain-like configurations that readily glide past one another. This gives water its fluid properties. In the vapor state (as found in the atmosphere), all of the hydrogen bonds between water molecules are broken and water exists as solitary molecules of gas.

Water and Precipitation

Water's chemistry makes it a special substance. But when water falls from the sky in any of the many forms of precipitation, it is not acting alone. Atmospheric particles serve as condensation nuclei onto which water vapor condenses. Condensation nuclei are often microscopic in size and are introduced into the atmosphere from the soil, plants, the ocean, and from natural and human-made combustion processes. Condensation nuclei are necessary to help the water droplets overcome the inhibiting effect of surface tension at the interface between the droplet and the air. As condensation continues, water molecules coalesce into cloud droplets, the precursors of raindrops and other forms of precipitation.

When the air in an air mass is below the freezing point of water and particles that serve as ice nuclei are present, ice crystals, rather than cloud droplets, may form. Ice crystals form in a number of hexagonal (six-sided) shapes and may develop into raindrops, snowflakes, or other forms of frozen precipitation. Under slightly subfreezing conditions, cloud droplets often exist in a supercooled liquid form rather than freezing immediately into ice crystals. This is because freezing nuclei that help the liquid water freeze—they do this by mimicking the molecular shape of an ice crystal—are rare in the atmosphere.

When cloud droplets, supercooled cloud droplets, or ice crystals form in sufficient numbers, they become visible as clouds.

The type, size, and shape of clouds depend on the nature of the cloud droplets or ice crystals and on existing atmospheric conditions. Clouds are most commonly classified according to their appearance, falling into one of three major subdivisions:

- *Cirrus* (hairlike): high, feathery clouds composed of ice crystals
- *Stratus* (layered): clouds composed of water droplets that form in layers
- *Cumulus* (pile or heap): detached clouds composed of water droplets having the appearance of a mound, dome or tower

Various prefixes and suffixes are added to these subdivision names to describe a particular cloud formation. For instance, when a cumulus cloud is producing rain or snow, the suffix "nimbus" is added, identifying the cloud as "cumulonimbus."

Precipitation develops within cumulonimbus and other storm clouds where sufficient numbers of cloud droplets, ice crystals, or supercooled cloud droplets have formed. When an ice crystal forms in a cloud of supercooled droplets, it rapidly grows at the expense of the surrounding droplets. (The reason for this will not be discussed here.) Now the ice crystal begins to fall and collects more and more droplets by collision. When the ice crystal passes below the freezing level in the atmosphere, it melts to form a rain drop. Ninety–five percent of the rain that falls to Earth begins as snow in this way in the upper reaches of the clouds.

If the raindrops then fall into air with a subfreezing temperature, they are supercooled and become freezing rain. When supercooled raindrops fall on cold surfaces, they freeze, producing a coating of ice. Sleet results when supercooled raindrops freeze in the air and reach the ground as small pellets of ice. Hailstones are ice pellets that are greater than 5 mm in diameter. They are formed in thunderstorms containing strong, persistent updrafts that carry the ice pellets up through colder air several times before they fall to Earth, giving the particles the opportunity to grow in size.

Precipitation and cloud formation, along with wind patterns and frontal zones, are extremely complex processes occurring within the atmosphere. Subtle variations in any number of conditions can turn a threatening or severe storm into a relatively harmless rain shower (see Reading 9, "The Inner Working of Severe Weather"). While varied and complex, weather phenomena—driven by the redistribution of heat energy in the atmosphere (see reading 5, "The Redistribution of Thermal Energy")—are all directly related to the unique properties of water.

The Inner Workings of Severe Weather

SCI LINKS. THE WORLD'S A CLICK AWAY

Topic: severe weather
Go to: www.scilinks.org
Code: PESM200

A vital function of weather forecasting is to provide timely warnings about the approach of dangerous weather. Accurately forecasting such events requires considerable information about developing weather systems. Much of this information is collected using observer networks, weather balloons, radars, and satellites. These data are compiled by the National Weather Service (NWS) and are made available to the public in numerous ways. But predicting weather changes accurately also requires understanding the mechanisms that underlie complex weather systems. Knowing how severe weather works is key for anticipating when it will appear. Let's consider thunderstorms, tornadoes, and hurricanes.

Severe Thunderstorms and Tornadoes

A thunderstorm is a convective cloud that produces lightning and thunder. Both these phenomena are linked to the formation of strong up and down drafts within the cloud itself, although the exact mechanism is not completely understood. During a thunderstorm, positive charges accumulate near the cloud top while negative charges build up in the cloud base. If the electric potential across the cloud becomes large enough to overcome the insulating properties of the air within the cloud, a huge electrical spark—called an arc—occurs. (Most lightning occurs within the clouds; only one spark in ten reaches the ground.) The arc results in a tremendous release of energy that superheats the surrounding air. This superheating, in turn, causes the air to expand explosively, resulting in a sonic boom: thunder. (For more on lightning and thunder, see Reading 10.)

Thunderstorms that occur in conjunction with a strong jet stream flow can produce tornadoes. In the United States, tornadoes occur most frequently in the late spring, although they can occur (and have occurred) in every region of the country and during every season of the year. Tornadoes appear as twisting gray funnels, cylinders, or ropes extending down from a cloud base. Two other identifying characteristics are an extremely loud, roaring noise—like a freight train—and an unusually intense period of lightning. Most tornadoes are fairly small—i.e. a few hundred meters in diameter—and usually are short lived, lasting only a minute or two in any one area. However, with wind speeds sometimes exceeding 190 knots (360 kph) and extremely low pressures, tornadoes can cause a great deal of damage as they pass. Cars, buildings, animals, and people may be thrown hundreds of meters. Water may be suctioned upward, leaving rivers, ponds, and small lakes dry. Many of the deaths that occur during tornadoes result from collisions with debris that is hurled through the air by the tornado's extremely strong winds.

Tropical Storms and Hurricanes

As devastating as tornadoes can be, historically they have been responsible for causing far less damage than hurricanes (called typhoons in the western Pacific). Hurricanes begin as atmospheric depressions—low pressure areas—over warm, tropical areas of the ocean. In these depressions, the air at the surface is heated and moistened by the warm ocean. As the warm air rises, it cools and loses its capacity to hold moisture. As a result, condensation and cloud formation begin, and latent heat is released into the atmosphere. A low pressure area develops beneath the rising warm air. This, in turn, causes an increase in the flow of air into the storm. Because of Earth's rotation, the incoming air is deflected to the right of its direction of motion (see Figure 1). (Of course, this rotation to the right—the Coriolis Effect—occurs only in the Northern Hemisphere; converging air in a Southern Hemisphere depression is deflected to the left of its direction of motion.)

The rotating winds converge within the low pressure area near the ocean's surface and are warmed. They also evaporate water from the ocean surface and carry it into the growing storm

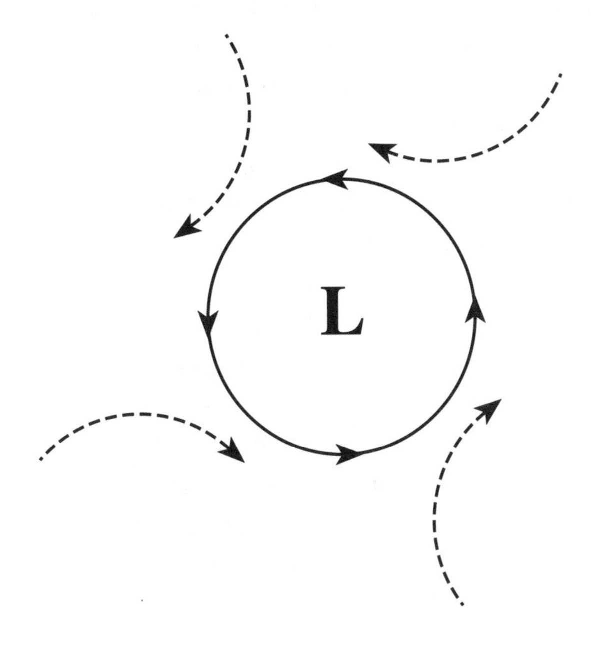

Figure 1. Converging winds are deflected into a circular pattern by the Coriolis Effect

system. Once inside the low pressure area, the moist air rises and the water vapor it contains begins to condense. The condensation process releases more latent heat, which warms the air further, and the cycle of storm development intensifies. As it continues, the pressure in the center of the depression further drops and the converging winds are drawn into the low pressure area at an increasingly faster rate. Precipitation begins in the region surrounding the central low pressure area. When maximum wind speeds within a tropical depression exceed 35 knots (66 kph), the storm is reclassified as a tropical storm and given an identifying name.

Most tropical storms (about 70 percent) continue to intensify while they remain over warm tropical waters, continuing the chain reaction described above. When maximum wind speeds within a tropical storm exceed 64 knots (118 kph), the storm is designated a hurricane. At these high speeds, the converging winds spin so rapidly that they are forced into a ring surrounding the central low pressure area. Within the central low—the eye of the hurricane—the winds are light, and there are few if any clouds. The eye is usually 20–40 km in diameter and is surrounded by a 20–50 km thick wall of intense thunderstorms, called the eye wall. The strongest winds of the storm—up to 150 knots (190–280 kph) in a mature hurricane—are found within the eye wall. Gusts there have been recorded in excess of 190 knots (360 kph)! Spiral cloud bands extend outward from the eye wall and rotate with the storm center. These bands consist of many individual thunderstorms.

As a hurricane and its associated thunderstorms move over the ocean, the winds create large waves, often the first actual warning sign of an approaching hurricane. When the storms approach a coastline, the winds and low pressure also cause a sharp increase in water level. This wind- and pressure-induced abnormal rise in sea level is called a storm surge. Storm surges cause considerable flooding in low-lying coastal areas. In fact, storm surges are responsible for most of the death and devastation caused by hurricanes.

Hurricanes can last more than a week. As they move over land or colder water, hurricanes lose their sources of energy. The eye disintegrates and the storm's overall intensity is reduced. However, before dissipating, hurricanes that move inland generally bring heavy rains, strong winds, and flooding. Some hurricanes have produced their worst flooding long after coming ashore.

Because hurricanes are fueled by warm ocean waters, "hurricane season" occurs only when the water temperature is high enough (above 26°C) to sustain the development of tropical storms. Temperatures in tropical waters usually reach this level in late summer and early fall in the northern hemisphere.

Severe Weather Watch/Warning System

In an effort to keep people informed about occurrences of severe weather, the NWS has devised a watch/warning system. Forecasters issue a severe thunderstorm or tornado watch when the conditions during the following few hours will be right for the formation of a severe thunderstorm or tornado over a given region. People within a watch area are advised to be on the lookout for dangerous weather. When a severe thunderstorm or a tornado is sighted by ground observers or by radar, a warning is issued, advising people within the surrounding area to take protective measures immediately. In many tornado-prone areas, communities have emergency preparedness plans for coping with such severe weather.

Weather satellites effectively track tropical storms and hurricanes over the entire Earth. When these storms move to within a few hundred kilometers of land, airplanes equipped with special instruments and ground radars also are used to track their movement and determine their characteristics. When a possibility exists that a hurricane will come ashore within the next 36 hours, a hurricane watch is issued by the NWS. If landfall is likely within the next 12 to 24 hours, a hurricane warning is issued. When these are announced, quick evacuation of people in low-lying areas is essential.

Thunderstorms, tornadoes, and hurricanes are only three kinds of severe weather patterns. Similarly dangerous are events such as lightning, blizzards, severe heat waves, and flash floods. Over the last 20 years, in fact, flash floods have been the most deadly severe weather problem in the United States. Watches and warnings need to be taken seriously if the risks to life and property are to be minimized.

Flash to Bang

- ***October 16, 1992, Houston:***
  35 football players and coaches at practice injured by lightning at 8:30 am during light drizzle.
- ***October 15, 1992, Chicago and northwest Indiana:***
  Six football players at practice, and two parents at a school's back door, were injured by lightning in three incidents.
- ***May 31, 1991, Colorado Springs:***
  Female soccer player killed and three others injured by lightning during a tournament.
- ***June 7, 1989, Orlando:***
  A 12-year-old boy taking shelter under a tree critically injured by lightning.

Introduction

Across the United States in 1992, as in earlier years, dozens of school children were killed or injured by lightning at outdoor extracurricular activities, on the way to or from school, or at recess.

Lightning is one of nature's most spectacular displays, yet it remains far from being well understood. Lightning is the most frequent important weather threat to personal safety during the thunderstorm season. If people were better educated about the dangers of lightning, practiced safety, and used common sense before, during, and after thunderstorms, lightning fatalities and injuries could be sharply reduced.

Deaths from Severe Weather

Lightning as a safety threat sometimes receives less attention than hurricanes, tornadoes, and flash floods. Yet since 1940, lightning has been the leading cause of weather related deaths (see Figure 1). Between 1940 and 1991, 8,316 people were killed by lightning in the United States, according to statistics from National Oceanic and Atmospheric Administration (NOAA).

Figure 1. Weather related deaths in the United States between 1940 and 1991

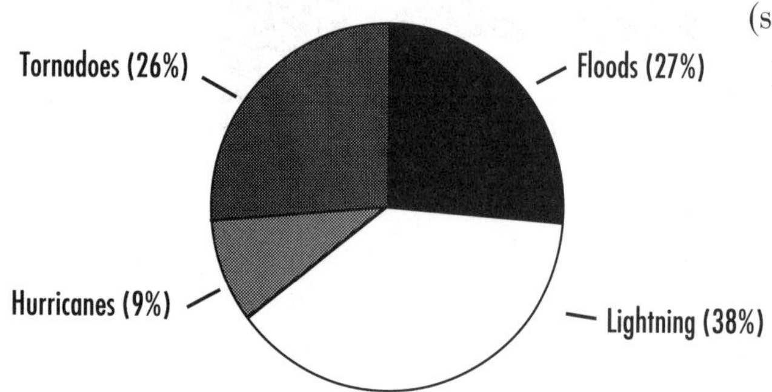

For the past 20 years, lightning ranked second only to flash floods in weather related deaths (see Figure 2). In addition, injuries from lightning are more than 2.5 times more numerous than lightning deaths, so the number of people affected by lightning every year is large. Tornadoes ranked third (69 killed per year), and hurricanes were fourth (17 per year). Tornadoes, hurricanes, and flash floods often result in multiple deaths and massive destruction of property, so they make headlines.

But lightning usually takes its victims one at a time, so it is not given the attention it requires as a common threat.

Lightning

To appreciate the scope of electrical activity and the threat from lightning, look at the over 15,000 lightning strikes to the ground in 6 hours from Iowa and Missouri to Illinois and Wisconsin, Indiana, and Michigan in Figure 3. This type of map is continuously produced by networks of electromagnetic sensors that chart cloud-to-ground lightning strikes for the continental United States and other countries. Such networks were developed for forest fire detection and utility company needs. But uses have expanded into such applications as refueling and baggage handling at airports, crowd management at golf tournaments, thunderstorm monitoring and forecasting by weather services, and understanding the nature of lightning itself. The episode in Figure 3 was related to an advancing cold front and associated thunderstorms. Two people were injured by lightning on this day at O'Hare Airport in Chicago.

Figure 2. Weather related deaths in the United States between 1972 and 1991

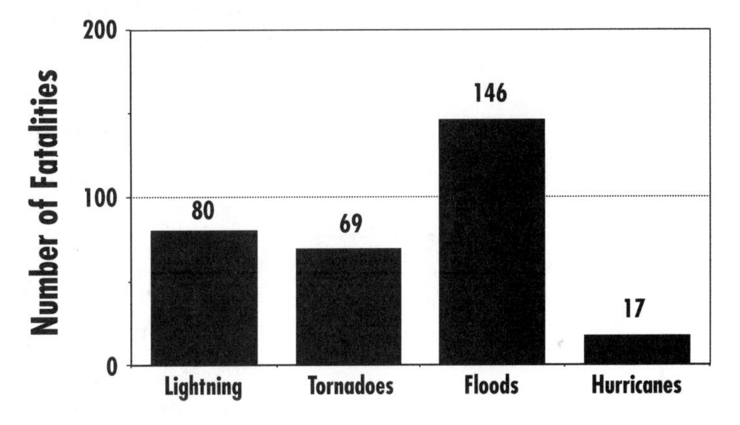

A satellite-based lightning sensor is being developed to be launched before the year 2000. It will plot in-cloud as well as cloud-to-ground flashes over a much larger portion of Earth than ground based networks.

Thunder

The distance to lightning from your location can be found using the fact that light travels enormously faster than sound. This difference leads to the "flash-to-bang" method:

- See the flash.
- Count the seconds to the bang of its thunder. Divide the number of seconds by five to give the approximate distance in miles from you to the lightning.

For example, suppose it takes 15 seconds between the time you see lightning and when you hear its thunder. Divide 15 by 5 to give a distance of about 3 miles from where lightning occurred to your location.

Thunder is produced when air immediately around the lightning channel is superheated, in less than a second, to 8,000–33,000 °C. When air is heated this quickly, it explodes. When lightning strikes nearby, the sound may be a loud bang, crack, or snap. There often follows a rumbling or growling sound caused by sounds from different heights along the channel at farther distances than the nearby strike point. The sound reaches our ears at varying times and may last for several seconds.

Thunder can be heard up to about 10 miles away (50 seconds from flash to bang). During heavy rain and wind, thunder can't be heard that far, but in a very quiet location, especially at night, it can be heard farther than 10 miles.

One recent study in Florida found that the average distance between successive ground strikes in the same storm was about 2–3 miles. Since this distance was an average, some flashes were within 2–3 miles, and the rest were beyond that distance. (Two to three miles corresponds to 10–15 seconds from flash to bang.) If you are within 2–3 miles of a flash, *the next flash can easily strike your location.* And remember that the 2–3 mile range is an *average*. Half of all subsequent lightning strikes fall beyond this range. (Lightning has been reported to land up to 16 km and farther away!) Knowing this makes clear why precautions are needed even when the flash-to-bang exceeds three miles.

Figure 3. Map of 15,422 lightning strikes to ground in a 6 hour period across the Midwest shown by plus, diamond and triangle symbols.

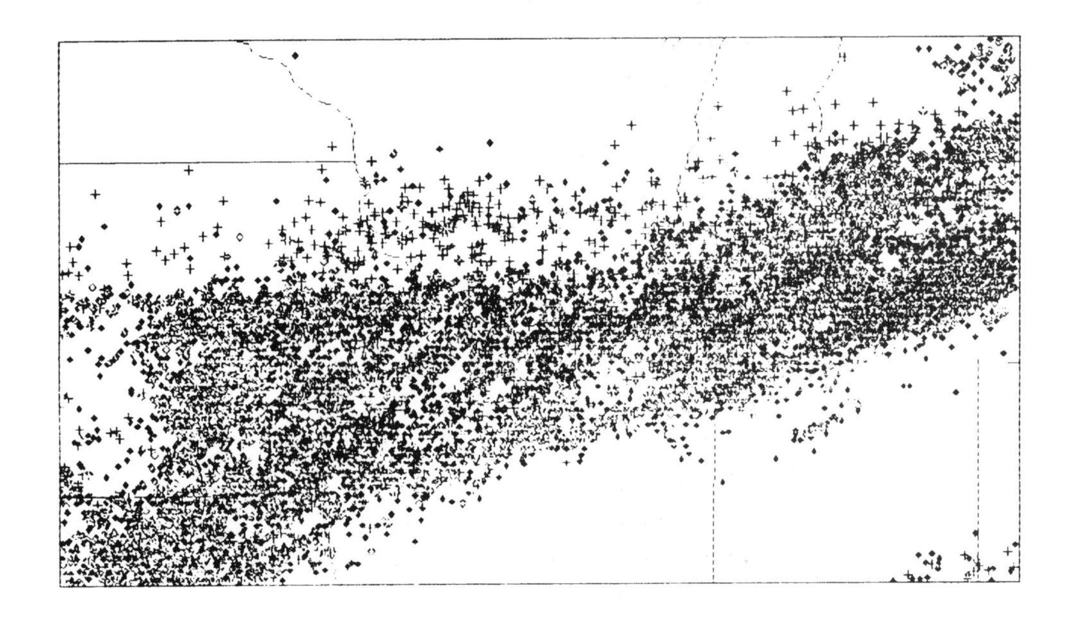

Lightning and People

National statistics show that males are killed by lightning 3 to 9 times more often than females, and mostly between the ages of 20 and 40. Most lightning fatalities occur between 10 am and 7 pm local time. People are outdoors most often during these hours, but thunderstorms occur most often in the afternoon and evening. Lightning deaths are most frequent from late spring through early autumn—July has the most fatalities (see Figure 4).

In the last five years, Florida had the most lightning deaths and injuries. In a recent study for central Florida, more lightning casualties occurred before or after the peak lightning activity. Compared to the middle of a lightning event, such a pattern may occur because the flash rate is lower, and it is not raining as hard.

In central Florida from 1983 to 1990, the most common lightning fatalities and injuries were, in order of occurrence:

1. Near or in the water
2. Near or under a tree
3. Near a vehicle, home, or building
4. On a golf course, ball field, or similar situation

In Colorado, more lightning casualties occur during climbing and hiking in the mountains than any other single category; in Florida, casualties during water sports are the most frequent.

Statistics show that agricultural lightning casualties are declining as people move to urban areas, and the recreational component is growing.

Lightning striking near a person, a side flash, can cause injury, blindness, deafness, or death. The side flash is a very important feature of the lightning hazard, because many casualties are not directly struck by the flash. The side flash hazard depends on many factors, especially the following:

- Distance to the ground strike point
- Soil conductivity (wet soil conducts better than dry soil)
- Strength of the electric current

Side flash deaths are most common outdoors, but also can happen indoors through telephones, electrical appliances, and water pipes connected to sinks, showers, or baths.

The risk of death or injury from a lightning strike may be greater than official data show. Recent research in Colorado indicates that deaths and injuries there are under reported. In addition, some deaths and injuries are not reported when they are caused indirectly by lightning, from a lightning-caused fire for example, but not directly resulting from a lightning strike.

The under reporting of lightning-caused deaths and injuries may be 30 percent or more.

About 20 percent of all people who are casualties of lightning are killed. It is estimated that the chances of any one person being a casualty of lightning is one in 600,000. Before his death due to other causes, Mr. Roy "Dooms" Sullivan, a park ranger from Virginia, was injured on seven different occasions by lightning.

Lightning and Property

Lightning strikes have damaged buildings, electrical transmitters, livestock, and aircraft electrical systems, and caused many forest and range fires. Any object has the potential to be struck by lightning, especially if it protrudes above the ground. Tall, metallic, isolated objects are especially vulnerable compared to other materials in the immediate surroundings.

When aircraft in flight are struck by lightning, the aircraft itself usually induces the flash. Typically, there is little or no damage to the flying aircraft. Commercial aircraft are struck by lightning once every 5,000 to 10,000 flight hours.

In 1967, Apollo 12 flight and ground crews suffered anxious moments when the launch triggered two lightning strikes to the craft, causing minor damage. On March 26, 1987, an unmanned Atlas-Centaur rocket with a satellite was destroyed when lightning struck it shortly after lift-off.

Over $100 million damage by lightning is reported annually in the United States, but the amount is much larger for several reasons:

- *Uninsured* minor but frequent damage is caused by direct or indirect lightning strikes to houses, electronics, trees, and many other objects. A location in the United States with an average lightning frequency, such as the Midwest or East, is estimated to be struck once within every one hundred-year period.
- *Protection* expenses are significant each year on such facilities as power lines and buildings.
- *Avoidance* costs are large (but hard to quantify) for construction crews who stop work in threatening weather, and recreation that is cancelled or delayed.

Figure 4. Monthly distribution of lightning deaths between 1959 and 1990 in the United States

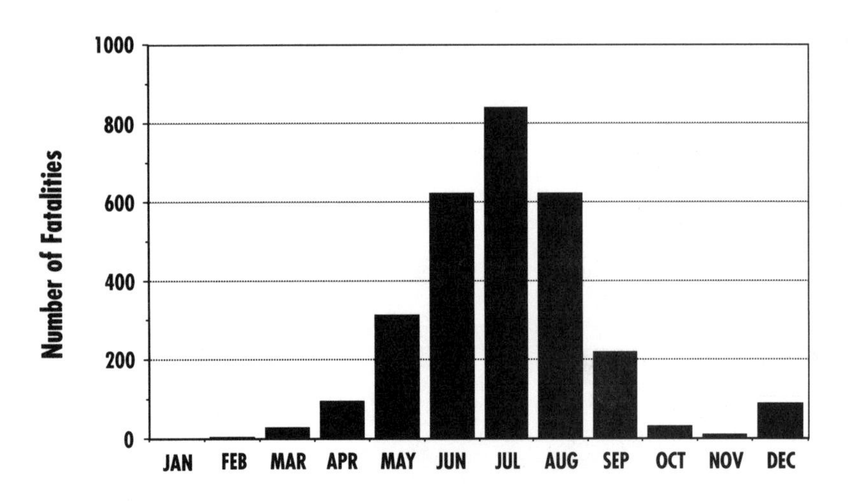

The Lightning Flash

Electrical activity has been observed on Venus, Jupiter, Saturn, and Uranus, and may occur on other planets. On Earth, lightning originates in thunderstorms and occasionally in clouds during volcanic eruptions. There is a beneficial effect of lightning: it produces fixed nitrogen, which is necessary for fertilization of plants.

At any given moment, 2,000 thunderstorms are estimated to be under way around the world, and lightning strikes the ground about 100 times each second, or 8 million times a day. The United States averages over 100,000 thunderstorms a year; China is second in the world with 85,000. We divide lightning into two general types:

- Cloud-to-ground lightning
- Cloud discharge or in-cloud lightning

Data from the new networks of electromagnetic sensors (these sensors produced the data for Figure 3) indicate that about 20 million flashes strike the ground in the United States annually. The United States has at least 100 million lightning flashes of both types every year.

Cloud-to-ground lightning originates from the base of a thunderstorm. These step leaders surge downward about 50 meters at a time, sometimes more in the horizontal than vertical, attempting to complete a channel with the ground. When the leader is close to the ground, one or more streamers reaches upward from such objects as grass, tall trees, and buildings. The last surge of a step leader, before completing the channel with ground, is usually vertical. The entire process of recurrent surges downward and streamers upward takes less than a second.

Once the channel is complete, a surge of electrical current from the ground moves upward along the channel, producing an upward-propagating bright luminosity. Negative charge, however, is being transferred to the ground. This discharge process is called a return stroke. After the first stroke, additional leaders can deposit negative charge down along the old channel and, as connection with the ground is established, another upward current surge occurs as the charge is brought to the ground. This process constitutes another return stroke. This can happen several times along the same channel.

The whole sequence of first and subsequent strokes is a flash (see Figure 5). A typical flash has two to four return strokes. But one cloud-to-ground flash near Cape Canaveral produced 26 return strokes and lasted over two seconds!

Figure 5. Diagram of cloud-to-ground lightning flash sequence

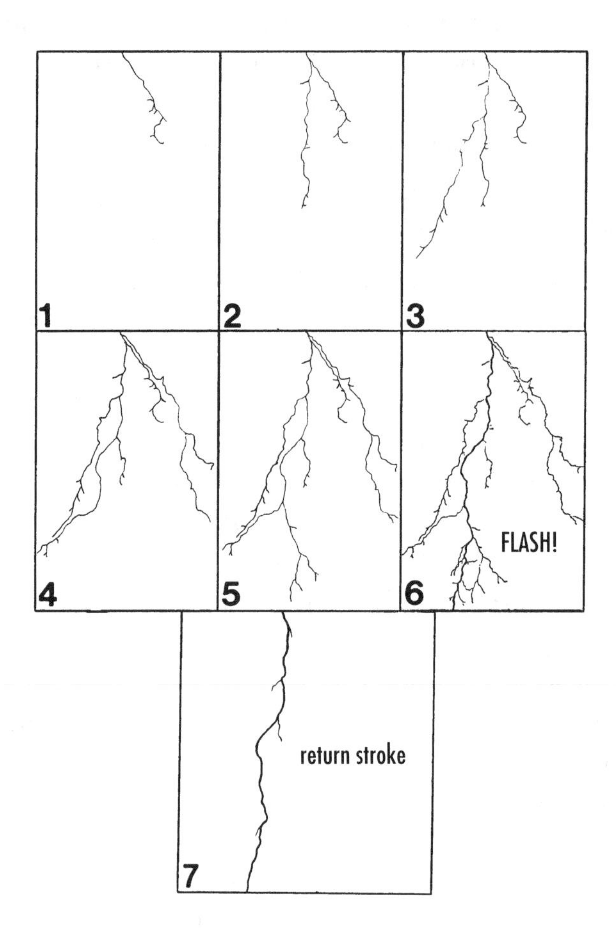

Because the leaders, streamers, and return strokes happen so fast, we cannot see luminosity propagate up the channel. But we can see a flickering of the luminosity corresponding to different return strokes. The actual diameter of a lightning channel is two to five centimeters. Note the terminology here:

- A *flash* consists of one or more *return strokes.* Flickering indicates more than one return stroke.
- A *ground strike* is a flash that hits the ground.

We use only the terms of flash, return stroke, and ground strike in this review—the word "bolt" has no clear definition for lightning.

A thunderstorm can produce several hundred megawatts of energy—comparable to the output of a small nuclear power plant. A flash has a billion volts of energy, with a peak current between 10,000 and 200,000 amperes. These amounts of energy account for the degree of damage to people and objects struck by lightning. The average flash could light a 100 watt bulb for more than three months. The brightness of a single flash is more than 10,000,000 light bulbs of 100 watts. It is no wonder that eye damage is common to people very near a strike. Unfortunately, harnessing this electrical power is not feasible. The flash would need to be where it is wanted, and it is so brief that no equipment could capture the tremendous surge without damage.

The Thunderstorm

Thunderstorms are usually like a dipole, where positive charges are found at the top, and negative charges at the base. The ground, normally negatively charged, reverses polarity under a thunderstorm and becomes positively charged. This reverse in polarity is caused by a strong negative charge at the base of the storm, inducing a positive charge below and around the storm on the ground. This positive charge follows the thunderstorm like a shadow until the storm dissipates. Charge separation in thunderstorms is produced in rapidly rising air interacting with precipitation (both ice particles and liquid drops) moving downward within the cloud. The stronger the convective activity, the stronger the electrical potential that is developed. Electrical charge in and around thunderstorms is very complicated and changes constantly on a scale of minutes.

The life cycle of lightning-producing storms can be as short as 30 minutes, although they often last for an hour (but they can continue for as long as 12 hours). Thunderstorms proceed through three stages.

- *Growth*: Towering cumulus clouds grow upward more rapidly than horizontally. A cumulus tower is visible as an isolated vertical column, or as a rounded bubble at the cloud top. A thunderstorm usually has several towers. Electrical activity in a cumulus tower usually starts when cloud-top temperatures reach -15 to -20 °C.
- *Mature*: Precipitation droplets form, and precipitation may be visible from the ground. The cloud is now a cumulonimbus, showing the smoother and streakier appearance of an icy top. It takes several minutes for precipitation from this tower to reach the surface. On the ground, then, you may see the following: lightning may occur before any rain reaches the ground. First ground flashes in a cloud usually occur at this stage, but a few in-cloud flashes may happen earlier. Strong updrafts and weaker downdrafts are next to each other at up to 80 kph or more, and are several kilometers wide and high. The mature stage is the most intense, and has the most precipitation and electrical activity.

- *Dissipation*: With time, a downdraft spreads out after it reaches the ground. Typically, this cooler, outflowing air cuts off the inflow of moist, warm air feeding the updraft, and the thunderstorm weakens. There may be a few last ground flashes in the early dissipating stage; they are especially dangerous because the storm may appear to have died. Finally, precipitation and lightning stop and the storm dies.

Some thunderstorms produce a great deal of lightning and little rain on the ground, especially in the western United States, where such situations lead to forest and range fires. Storms in a moist subtropical environment may produce heavy rain and only a few flashes. Cloud-to-ground lightning can occur outside heavy precipitation, so taking shelter *only* when heavy rain is falling is *not* an adequate precaution from lightning.

Some Precautions

Procedures for avoiding lightning are summarized in safety pamphlets available from your nearest National Weather Service/NOAA and American Red Cross offices. Review them and *use common sense* before, during, and immediately after the strongest parts of thunderstorms. Even if no lightning has occurred, but conditions for lightning are favorable, take the proper precautions and practice lightning safety rules (see box on opposite page). Pay much more attention to the lightning threat than to the rain.

Once the storm has begun, remember the flash-to-bang method:

- When you see the flash,
- Count the seconds to the bang of its thunder, and divide by five.

This gives the approximate distance in miles to the lightning from you. The average distance from one flash to the next in the same storm is 2–3 miles—10 to 15 seconds flash-to-bang. *Do not wait until a flash is close to you.* Remember, many strikes occur beyond this average distance too! Use good judgment for yourself, your family, friends, students, and others around you.

By James Vavrek (Spohn School Weather Station, Hammond, IN), Ronald L. Holle (National Severe Storms Laboratory, Norman OK), Jim Allsopp (National Weather Service, Romeoville, IN)

Reprinted from *The Earth Scientist*, Fall 1993, pp. 3–8.

Avoid dangerous lightning situations!

1. Plan Ahead!

• Be aware of other storms in your area. If today has active thunderstorms or if storms are predicted in your area, don't be caught where you can't take shelter on short notice. Watch for signs of a thunderstorm; they can grow quickly!

• Give yourself time to reach a safe place before lightning is an immediate threat. Storms can grow from the small towering cumulus stage into a lightning producer *in less than half an hour*.

• Designate a spotter to watch for the threat of lightning. Decide on rules for stopping whatever you and your group are doing; and decide in advance where to take shelter when it is necessary.

2. Don't be the highest object.

• Avoid standing in an open field or parking lot, on a bicycle along an open road, on top of a mountain or rock outcrop, on a boat in the open water.

• Don't be connected to or stand under anything taller than its surroundings: big trees, antennas, and towers. Don't touch anything connected to power lines, phone lines, cables, or plumbing coming into a building from the outside.

• Crouch on the balls of your feet, with your head down.

3. Always Use Common Sense

• Go inside a sturdy building or into a vehicle with a solid metal top. Do not be in contact with any metal on the building or vehicle.

• Follow your safety plan, regardless of the stage of the game, the hike, or the fishing. You can always return to it *after* the storm has passed.

*sci*LINKS

Project Earth Science: Meteorology brings you *sci*LINKS, a creative project from NSTA that blends the best of the two main educational "drivers"—textbooks and telecommunications—into a dynamic new educational tool for all children, their parents, and their teachers. This *sci*LINKS effort links specific textbook and supplemental resource locations with instructionally rich Internet resources. As you and your students use *sci*LINKS, you'll find rich new pathways for learners, new opportunities for professional growth among teachers, and new modes of engagement for parents.

In this *sci*LINKed text, you will find an icon near several of the concepts you are studying. Under it, you will find the *sci*LINKS URL (http://www.scilinks.org/) and a code. Go to the *sci*LINKS Web site, sign in, type the code from your text, and you will receive a list of URLs that are selected by science educators. Sites are chosen for accurate and age-appropriate content and good pedagogy. The underlying database changes constantly, eliminating dead or revised sites or simply replacing them with better selections. The ink may dry on the page, but the science it describes will always be fresh.

The selection process involves four review stages:

1. First, a cadre of undergraduate science education majors searches the World Wide Web for interesting science resources. The undergraduates submit about 500 sites a week for consideration.
2. Next, packets of these Web pages are organized and sent to teacher-Webwatchers with expertise in given fields and grade levels. The teacher-Webwatchers can also submit Web pages that they have found on their own. The teachers pick the jewels from this selection and correlate them to the *National Science Education Standards*. These pages are submitted to the *sci*LINKS database.
3. Then scientists review these correlated sites for accuracy.
4. Finally, NSTA staff approves the Web pages and edits the information provided for accuracy and consistent style.

Who pays for *sci*LINKS? *sci*LINKS is a free service for textbook and supplemental resource users, but obviously someone must pay for it. Participating publishers pay a fee to NSTA for each book that contains *sci*LINKS. The program is also supported by a grant from the National Aeronautics and Space Administration (NASA).

Resources

This Resources list was compiled by the staff, consultants, and participants of Project Earth Science, and by the NSTA Special Publications editors. It is not meant to be a complete representation of resources in meteorology, but will assist teachers in further exploration of this subject. The entries are subdivided into the folowing categories:

Activities, Books, and Curriculum Projects. This category includes complete curricula, multidisciplinary units, and collections of hands-on activities.

Reference Materials. Reference books and CD-ROMS are included in this category.

Audiovisual Materials. Media materials listed in this section include videotapes, videodisks, posters, and transparencies.

Instructional Aids. Included in this category are charts, games, photographs, and posters.

Activities, Books, and Curriculum Projects

University of California, Berkeley
GEMS
Lawrence Hall of Science, #5200
Berkeley, CA 94720-5200
(510) 642-7771
http://www.lhs.berkeley.edu/GEMS/

Acid Rain
Grades 6–10, 1990, ISBN# 0-912511-74-5
Part of the *Great Explorations in Math and Science* (GEMS) series. Activities investigate acid rain and its effects on living systems. (176 pages)

National Science Teachers Association
1840 Wilson Blvd.
Arlington, VA 22201-3000
(800) 722-NSTA
http://www.nsta.org/scistore/

Earth-Ocean-Atmosphere Explorer
Grades 6–College, 1998, #MS290X1
A collection of resources presenting basic geology, oceanography, and meteorology that includes a CD-ROM with narration, digital video, and interactive exercises and projects, as well as a geology lab manual and teacher's guide. (Mac/Win CD-ROM, 160-page lab manual, 60-page teacher's guide)

National Science Teachers Association
1840 Wilson Blvd.
Arlington, VA 22201-3000
(800) 722-NSTA
http://www.nsta.org/scistore/

Flood!
Grades 7–10, 1996, #OP460X5 (student), OP461X5 (teacher)
Part of the *Event-Based Science Modules* by Russell G. Wright; this book uses a recent major event and a combination of reading, writing, and hands-on activities to emphasize the interdisciplinary nature of science. Makes use of research and news stories that accompanied the event. Teacher guide comes with a video. (76 pages)

National Science Teachers Association
1840 Wilson Blvd.
Arlington, VA 22201-3000
(800) 722-NSTA
http://www.nsta.org/scistore/

Forecasting the Future
Grades 6–10, 1996, #PB118X
Developed jointly by NSTA and the Scripps Institution of Oceanography, this volume has 14 activities and more than 40 extension exercises to help students understand what global climate change means for the future. (160 pages)

University of California, Berkeley
GEMS
Lawrence Hall of Science, #5200
Berkeley, CA 94720-5200
(510) 642-7771
http://www.lhs.berkeley.edu/GEMS/

Global Warming & The Greenhouse Effect
Grades 7–10, 1992, ISBN# 0-912511-75-3
Part of the *Great Explorations in Math and Science* (GEMS) series. Activities investigate global warming and the greenhouse effect. (176 pages)

The Handy Weather Answer Book
Grades 6–12, 1997, #OP395X2
Answers to more than 1,000 frequently asked questions pertaining to weather-related topics. (432 pages)

National Science Teachers Association
1840 Wilson Blvd.
Arlington, VA 22201-3000
(800) 722-NSTA
http://www.nsta.org/scistore/

Hurricane!
Grades 7–10, 1995, #OP460X3 (student) OP461X3 (teacher)
Part of the *Event-Based Science Modules* by Russell G. Wright; this book uses a recent major event and a combination of reading, writing, and hands-on activities to emphasize the interdisciplinary nature of science. Makes use of research and news stories that accompanied the event. Teacher guide comes with a video. (60 pages)

National Science Teachers Association
1840 Wilson Blvd.
Arlington, VA 22201-3000
(800) 722-NSTA
http://www.nsta.org/scistore/

Investigating Air
Grades 9–12, 1998, #PB148X
Students use current online data and generate their own data to explore air quality issues such as increases in atmospheric carbon dioxide, acid deposition, ozone, and visibility. Developed by NSTA. (78 pages)

National Science Teachers Association
1840 Wilson Blvd.
Arlington, VA 22201-3000
(800) 722-NSTA
http://www.nsta.org/scistore/

Looking at Weather
Grades 4–8, 1991, by D. Suzuki, #K-0-471-54047-1
Describes the changes in weather, how weather affects people's lives, and how people affect weather. Includes activities. (96 pages)

John Wiley & Sons, Inc.
605 Third Avenue
New York, NY 10158-0012
(800) 850-6000
http://www.wiley.com

Oceans and Atmosphere
Grades 6–12, 1992, ISBN# 0-835-90385-0 (student), 0-835-90390-7 (teacher)
Part of the *Science Workshop Series*, these supplemental activities and experiments focus on oceans and the atmosphere.

Globe Fearon
4350 Equity Drive
PO Box 2649
Columbus, OH 43216
(800) 848-9500

Pollution GeoKit
Grades 5–9, 1996, #Y90559 (entire kit)
Activities focus on contamination of air, water, and land. Includes teacher's guide, transparencies, maps, posters, handouts, videos, and *National Geographic* articles. Can be purchased as an entire kit, or as individual elements.

National Geographic Society
1145 17th Street, NW
Washington, DC 20036-4688
(800) 368-2728
http://www.ngstore.com/

Pressure
Grades 7–12, 1992
Part of the *Task Oriented Physical Science* (TOPS) program, this module includes 32 activities that illustrate the fundamental concepts of pressure.

TOPS Learning Systems
10970 S. Mulino Road
Canby, OR 97013
(888) 773-9755
http://topscience.org/

Studies in Weather and Climate
Grades 9–college, 1991, by P. Suckling
Supplementary meteorology exercises provide extensive practice with maps, charts, and graphs. Use of a calculator is strongly recommended for some activities. (206 pages)

Contemporary Publishing
Company of Raleigh
6001-101 Chapel Hill Rd.
Raleigh, NC 27607
(919) 851-8221

Tornado!
Grades 7–10, 1996, #OP460X8 (student) OP461X8 (teacher)
Part of the *Event-Based Science Modules* by Russell G. Wright; this book uses a recent major event and a combination of reading, writing, and hands-on activities to emphasize the interdisciplinary nature of science. Makes use of research and news stories that accompanied the event. Teacher kit includes video. (64 pages)

National Science Teachers Association
1840 Wilson Blvd.
Arlington, VA 22201-3000
(800) 722-NSTA
http://www.nsta.org/scistore/

Water & Air
Grades 6–12, 1995, ISBN# 0-835-90745-7 (student),
0-835-90746-5 (teacher)
Part of the *Environmental Science* series. The teacher resource manual provides answers and background information, reproducible report sheets and test masters, activities and interdisciplinary connections. (96 pages)

Globe Fearon
4350 Equity Drive
PO Box 2649
Columbus, OH 43216
(800) 848-9500

Weather GeoKit
Grades 5–9, 1996, #Y90558 (entire kit)
Activities allow students to predict local weather, learn about forecasting, weather maps, layers of the atmosphere, global winds, and pressure systems. Includes teacher's guide, transparencies, maps, posters, handouts, videos, and *National Geographic* articles. Can be purchased as an entire kit, or as individual elements.

National Geographic Society
1145 17th Street, NW
Washington, DC 20036-4688
(800) 368-2728
http://www.ngstore.com/

Weather Workstation
Grades 6–College, 1998, #MS290X2
A tutorial of 25 lessons, including lessons on the fundamentals of weather, hurricanes, tornadoes, severe winter storms, climate change, and ozone. (Mac/Win CD-ROM)

National Science Teachers Association
1840 Wilson Blvd.
Arlington, VA 22201-3000
(800) 722-NSTA
http://www.nsta.org/scistore/

What Air Can Do
Grades K–2, 1983, #U04826
Learn about wind, how it moves things, and why all plants and animals need air to survive. Contains teacher's guide, read-along narration cassette, 30 student booklets, and reproducible activity sheets.

National Geographic Society
1145 17th Street, NW
Washington, DC 20036-4688
(800) 368-2728
http://www.ngstore.com/

Why Does it Rain?
Grades K–2, 1983, #U04822
Examine the water cycle including evaporation, cloud formation, and precipitation. Contains teacher's guide, read-along narration cassette, 30 student booklets, and reproducible activity sheets.

National Geographic Society
1145 17th Street, NW
Washington, DC 20036-4688
(800) 368-2728
http://www.ngstore.com/

Reference Materials

National Geographic Society
1145 17th Street, NW
Washington, DC 20036-4688
(800) 368-2728
http://www.ngstore.com/

Earth's Climate
Grades 4–9, 1998, #U86045
How altitude, land forms, and ocean proximity create tropical, temperate, and polar regions; climate shifts, and phenomena such as the ice ages and El Niño are all covered in this PictureShow CD-ROM. Includes a user's guide, class activities and student information, English or Spanish narration and text, and library catalog cards. (see also *Earth's Climate* transparencies under Instructional Aids below)

National Geographic Society
1145 17th Street, NW
Washington, DC 20036-4688
(800) 368-2728
http://www.ngstore.com/

Introduction to Weather
Grades 4–9, 1998, #U86043
Everything from the interaction of air, weather and the sun's energy as it affects weather, how weather is forecast, and severe weather events such as tornadoes, hurricanes, and thunderstorms are all covered in this PictureShow CD-ROM. Includes a user's guide, class activities and student information, English or Spanish narration and text, and library catalog cards. (see also *Introduction to Weather* transparencies under Instructional Aids below)

Operation: Weather Disaster
Grades 5–12, #718007 (Windows) #717991 (Macintosh)
Weather games and puzzles using a factual database allows students to test their weather knowledge, logic, and problem-solving skills individually or in teams. Comes with a teacher's guide.

The Discovery Channel School
PO Box 67027
Florence, KY 41022
(888) 892-3484
http://school.discover.com/

The Weather Book
Grades 7–college, 1992, #OP179X
A new approach to understanding the weather and how it occurs. Developed by *USA Today*, with well-designed graphics. (212 pages)

National Science Teachers Association
1840 Wilson Blvd.
Arlington, VA 22201-3000
(800) 722-NSTA
http://www.nsta.org/scistore/

The Weather Companion: An Album of Meteorological History, Science, Legend, and Folklore
General, 1988, by G. Lockhart
For naturalists, sportsmen, gardeners and sky watchers of all ages, here is a compendium of fascinating weather lore and facts, from myths to current research. (240 pages)

John Wiley & Sons, Inc.
605 Third Avenue
New York, NY 10158-0012
(800) 850-6000
http://www.wiley.com/

The Weather Wizard's Cloud Book
Grades 6–12, 1989, by L.D. Rubin and J. Duncan
A book about weather forecasting using cloud observations. (88 pages)

Algonquin Books
c/o Workman Publishing, Inc.
708 Broadway
New York, NY 10003
(212) 254-5900

Audiovisual Materials

After the Hurricane
Grades 4–9, 1993, #Y58012
Learn about the event and aftermath of one of the most destructive hurricanes ever to hit the United States. (27 minute video)

National Geographic Society
1145 17th Street, NW
Washington, DC 20036-4688
(800) 368-2728
http://www.ngstore.com/

Atmosphere: On the Air
Grades 4–9, 1993, #Y51580
Join the staff of a student-produced radio science show. Lean how the atmosphere protects the Earth and how it supports life; observe wind, clouds, and precipitation. (21 minute video)

National Geographic Society
1145 17th Street, NW
Washington, DC 20036-4688
(800) 368-2728
http://www.ngstore.com/

The Climate Puzzle
Grades 6–12, 1986
Number three in the *Planet Earth* series. Looks at modern changes in climate. (60 minute video)

Corporation for Public Broadcasting
The Annenburg/CPB Project
PO Box 2345
South Burlington, VA 05407-2345
(800) 532-7637

Investigating Global Warming
Grades 7–9, 1997, #Y52571
Students watch scientists examine ice cores, generate computer models, and simulate a global-warming scenario in Biosphere II. (22 minute video)

National Geographic Society
1145 17th Street, NW
Washington, DC 20036-4688
(800) 368-2728
http://www.ngstore.com/

Ozone: Protecting the Invisible Shield
Grades 9–College, 1994, #Y51616
Learn about the history of the ozone "hole" and the chemistry behind the phenomenon, as well as see how scientists, citizens, industry, and governments are working to halt further destruction of the ozone layer. (25 minute video)

National Geographic Society
1145 17th Street, NW
Washington, DC 20036-4688
(800) 368-2728
http://www.ngstore.com/

Telling the Weather
Grades 4–9, 1996, #Y50595
Students travel in the eye of a hurricane, see tornadoes, and learn about the latest equipment used in forecasting, as well as learn about what creates weather, how warm and cold fronts form, what conditions cause precipitation. (20 minute video)

National Geographic Society
1145 17th Street, NW
Washington, DC 20036-4688
(800) 368-2728
http://www.ngstore.com/

Weather: Come Rain, Come Shine
Grades 6–College, 1983, #Y51241
Through animation and live-action footage, students will learn about the global weather machine—the complex interaction of sun, air, and water that creates weather. (22 minute video)

National Geographic Society
1145 17th Street, NW
Washington, DC 20036-4688
(800) 368-2728
http://www.ngstore.com/

When Lightning Strikes
Grades 4–9, 1997, #Y52664
Animation illustrates how lightning occurs, and interviews with survivors of lightning strikes will enlighten students about this deadly natural force. (17 minute video)

National Geographic Society
1145 17th Street, NW
Washington, DC 20036-4688
(800) 368-2728
http://www.ngstore.com/

Where Storms Begin
Grades 4–9, 1999, #Y52813
Dramatic footage of storms, descriptions of their origins, climatic belts, colliding air masses, and more are covered in this video. (23 minute video)

National Geographic Society
1145 17th Street, NW
Washington, DC 20036-4688
(800) 368-2728
http://www.ngstore.com/

Wonders of Weather
Grades 6–12, #71774
Covers tornadoes, hailstones, how oceans influence weather and the latest hurricane-tracking technology. (45 minute video)

The Discovery Channel School
PO Box 67027
Florence, KY 41022
(888) 892-3484
http://school.discovery.com/

Instructional Aids

National Science Teachers Association
1840 Wilson Blvd.
Arlington, VA 22201-3000
(800) 722-NSTA
http://www.nsta.org/scistore/

Clouds
Grades K–5, #MS212X
Poster of cumulus, cirrus, and stratus clouds shown in four-color detail. (43 X 56 cm poster)

National Science Teachers Association
1840 Wilson Blvd.
Arlington, VA 22201-3000
(800) 722-NSTA
http://www.nsta.org/scistore/

The Cloud Charts
General, 1988, PS013X
A series of three full-color posters developed by NSTA, Weatherworks, and the National Weather Association, that provide photographs and information about cloud types, weather associated with them, and weather-related optical phenomena. (Set of three, 30.5 X 61 cm posters)

National Geographic Society
1145 17th Street, NW
Washington, DC 20036-4688
(800) 368-2728
http://www.ngstore.com/

Earth's Climate
Grades 4–9, 1998, #U86329
Transparencies on how latitude, altitude, and ocean proximity create climate zones. (See also *Earth's Climate* CD-ROM under Books, Booklets, and CD-ROMs above)

Introduction to Weather
Grades 4–9, 1998, #U86328
Transparencies on the interaction of air, weather, and the sun's energy to create weather systems, destructive thunderstorms, tornadoes, and hurricanes. (See also *Introduction to Weather* CD-ROM under Books, Booklets, and CD-ROMs above)

National Geographic Society
1145 17th Street, NW
Washington, DC 20036-4688
(800) 368-2728
http://www.ngstore.com/

Ozone
Grades 2–12, 1995, #U81924
Poster set, with teacher's guide, of images and facts relating to the Ozone layer.
(Set of three, 56x 85 cm posters)

National Geographic Society
1145 17th Street, NW
Washington, DC 20036-4688
(800) 368-2728
http://www.ngstore.com/

Pollution
Grades 2–12, 1995, #U81918
Poster set, with teacher's guide, of images and facts relating to worldwide pollution.
(Set of three, 56 x 85 cm posters)

National Geographic Society
1145 17th Street, NW
Washington, DC 20036-4688
(800) 368-2728
http://www.ngstore.com/

World Wide Weather
Grades 4–9, 1995, PS032X2
This striking poster shows winds, precipitation, storms, weather instruments, and map symbols. (56 X 89 cm poster)

National Science Teachers Association
1840 Wilson Blvd.
Arlington, VA 22201-3000
(800) 722-NSTA
http://www.nsta.org/scistore/

Activity #	Subject Matter and Content	Scientific Inquiry	Unifying Concepts and Processes
Activity 1	local weather observation and connection to larger weather patterns	how we know what we know in meteorology	nature of systems
Demo 2	origins of the Earth's atmosphere		evidence and models
Activity 3	existence and effects of air pressure	applying existing knowledge to new situations	evidence, models, and predictions
Activity 4	Earth's atmosphere has several gaseous components	relationship between explanation and evidence	gather, analyze, and interpret data
Activity 5	Earth's atmosphere has particulate components	relationship between explanation and evidence	gather, analyze, and interpret data
Activity 6	energy from the Sun affects temperature and behavior of the atmosphere	relationship between explanation and evidence	evidence, models, and predictions
Activity 7	absorption and reflection of light energy	relationship between explanation and evidence	evidence, models, and predictions
Activity 8	relationship between temperature and air pressure	interaction between concepts within a system	measuring changes within a system; evidence and prediction
Activity 9	relationship between air pressure and wind	relationship between explanation and evidence	evidence, models, and predictions
Demo 10	modeling the hydrologic cycle		evidence, models, and predictions
Activity 11	ways that water moves through the environment		systems, order, and organization
Activity 12	conditions required for cloud formation	relationship between explanation and evidence	change within a system
Activity 13	dew point—the amount of water vapor in the air	relationship between explanation and evidence	role of water in the Earth's system; change within a system
Activity 14	phase changes of water	relationship between explanation and evidence	change within a system
Activity 15	building a tool to measure relative humidity	relationship between explanation and evidence	change within a system
Activity 16	types of clouds indicate weather changes	relationship between explanation and evidence	evidence, models, and predictions
Activity 17	interpreting weather maps	tools for thinking critically and logically about data	models and predictions
Activity 18	how hail forms	relationship between explanation and evidence	change within a system
Activity 19	components and path of a hurricane	integration of multiple science processes	evidence, models, and predictions

Technology	Personal/Social Perspectives	Historical Context	Activity Name
			Weather Watch
			Making Gas
technology of well design		history of well design	The Pressure's On
designing and building a measurement tool	greenhouse effect, acid rain		The Percentage of Oxygen in the Atmosphere
	air pollution issues		It's in the Air
building and modifying a model			Why Is It Hotter at the Equator than at the Poles?
building and modifying a model	temperature differences in the built environment		Which Gets Hotter: Light or Dark Surfaces?
building a model and using it to make predictions			Up, Up, and Away!
building and using a measurement tool		history of rocketry	Why Winds Whirl Worldwide
building and observing a hydrologic tool	water use and water pollution		Recycled Water: The Hydrologic Cycle
	water use and water pollution		Rainy Day Tales
building and modifying a cloud chamber	smog formation		A Cloud in the Jar
building and using a measurement tool			Just Dew it!
building and using a measurement tool			Let's Make Frost
designing and building a measurement tool	personal health and climate		It's All Relative
			Moving Masses
	weather as a natural hazard		Interpreting Weather Maps
building and using a measurement tool	weather as a natural hazard		Hail in a Test Tube
	weather as a natural hazard	tracing the history of a hurricane	Chasing Hurricane Andrew

Poetry Credits

Robin Page, "Weather Sins" from *Weather Forecasting the Country Way* by Robin Page. Copyright © 1977 by Robin Page. Reprinted by permission of Summit Books, a division of Simon & Schuster, Inc.

Alan Lightman, "First Rainfall" excerpted from *Songs from Unsung Worlds* by Bonnie Bilyeu Gordon (Boston: Birkbauser). Copyright © 1985 by American Association for the Advancement of Science. Reprinted by permission.

"This Planet Earth" excerpted from *This Island Earth* edited by Oran W. Hicks, Scientific and Technical Information Division, Office of Technology Utilization, NASA, Washington, DC. Copyright © 1970, p. 36.

Myra Cohn Livingston, "Smog" and "Coming Storm" from *Sky Songs* by Myra Cohn Livingston. (Holiday House) Copyright © 1984 by Myra Cohn Livingston. Reprinted by permission of Marian Reiner for the author.

Langston Hughes, "Sun Song" from *Selected Poems* by Langston Hughes. Copyright © 1927 by Alfred A. Knopf, Inc. and renewed 1955 by Langston Hughes. Reprinted by permission of Alfred A. Knopf, Inc.

Sara Teasdale, "Wind Elegy" reprinted by permission of Macmillian Publishing Company from *Collected Poems of Sara Teasdale*. Copyright © 1926 by Macmillan Publishing Company, renewed 1954 by Mamie T. Wheless.

Christina Rossetti, "Who Has Seen the Wind?" from *Poems Children will Sit Still For*, by Christina Rossetti. Copyright © 1969.

Robert Frost, "Atmosphere" and "Our Hold on the Planet" from *The Poetry of Robert Frost*, edited by Edward Connery Lathem. Copyright © 1940, 1942, 1956 by Robert Frost. Copyright © 1970 by Lesley Frost Ballantine. Copyright © 1928, 1968, 1969 by Henry Holt and Company, Inc. Reprinted by permission of Henry Holt and Company, Inc.

Robert Frost, "It Bids Pretty Fair" from *The Poetry of Robert Frost*, edited by Edward Connery Lathern. Copyright © 1942, 1962 by Robert Frost. Copyright © 1970, 1975 by Lesley Frost Ballantine. Copyright © 1934, 1947, 1969 by Henry Holt and Company, Inc. Reprinted by permission Henry Holt and Company, Inc.

Langston Hughes, "April Rain Song" from *The Dream Keeper and Other Poems* by Langston Hughes. Copyright © 1932 by Alfred A. Knopf, Inc. and renewed 1960 by Langston Hughes. Reprinted by permission of the publisher.

Francisco Alarcon, "Clouds" and "Calendar Keepers" from *Snake Poems: An Aztec Invocation* by Francisco X. Alarcon. Copyright © 1992. Reprinted by permission of Chronicle Books.

Langston Hughes, "Snail" from *Selected Poems by Langston Hughes*. Copyright © 1947 by Langston Hughes. Reprinted by permission of Alfred A. Knopf, Inc.

Robert Louis Stevenson, 'Wintertime" from *A Child's Garden of Verses* by Robert Louis Stevenson, published by Longmans, Green (London, 1885).

Carl Sandburg, "Fog" from *Chicago Poems* by Carl Sandburg. Reprinted by permission of Harcourt Brace and Company.

Seamus Heaney, "Hailstones" from *The Haw Lantern* by Seamus Heaney. Copyright © 1987 by Seamus Heaney. Reprinted by permission of Farrar, Straus & Giroux, Inc.

Archibald MacLeish, "Hurricane" from *Collected Poems 1917-1982* by Archibald MacLeish. Copyright © 1985 by the Estate of Archibald MacLeish. Reprinted by permission of Houghton Mifflin Co. All rights reserved.